GROUND STUDIES FOR PILOTS

Volume 1

Volumes in the Series

Volume 1: Radio Aids
Volume 2: Plotting and Flight Planning
Volume 3: Navigation General and Instruments
Volume 4: Meteorology
and
Aviation Law for Pilots

GROUND STUDIES FOR PILOTS

Fifth Edition
Volume 1

RADIO AIDS

R. B. Underdown
FRMetS MRAeS
formerly Director of Ground Training and latterly Principal, College of Air Training, Hamble

Blackwell
Science

Fifth edition

Blackwell Science Ltd
Editorial Offices:
Osney Mead, Oxford OX2 0EL
25 John Street, London WC1N 2BL
23 Ainslie Place, Edinburgh EH3 6AJ
350 Main Street, Malden
MA 02148 5018, USA
54 University Street, Carlton
Victoria 3053, Australia
10, rue Casimir Delavigne
75006 Paris, France

Other Editorial Offices:

Blackwell Wissenschafts-Verlag GmbH
Kurfürstendamm 57
10707 Berlin, Germany

Blackwell Science KK
MG Kodenmacho Building
7-10 Kodenmacho Nihombashi
Chuo-ku, Tokyo 104, Japan

First published in Great Britain by
Crosby Lockwood and Son Ltd 1970
Second edition 1974
Third edition, in three volumes, 1979
Fourth edition published by Collins
Professional and Technical Books 1986
Fifth edition published by Blackwell
Science 1993
Reprinted 1995, 1996, 1998, 1999

Set by Best-set Typesetter Ltd, Hong Kong
Printed and bound in Great Britain by
Biddles Ltd, Guildford, Surrey

DISTRIBUTORS

Marston Book Services Ltd
PO Box 269
Abingdon
Oxon OX14 4YN
(*Orders:* Tel: 01235 465500
Fax: 01235 465555)

USA
Blackwell Science, Inc.
Commerce Place
350 Main Street
Malden, MA 02148 5018
(*Orders:* Tel: 800 759 6102
781 388 8250
Fax: 781 388 8255)

Canada
Login Brothers Book Company
324 Saulteaux Crescent
Winnipeg, Manitoba R3J 3T2
(*Orders:* Tel: 204 837-2987
Fax: 204 837-3116)

Australia
Blackwell Science Pty Ltd.
54 University Street
Carlton, Victoria 3053
(*Orders:* Tel: 03 9347 0300
Fax: 03 9347 5001)

A catalogue record for this title is available from the British Library

ISBN 0–632–03601–X

Library of Congress
Cataloging-in-Publication Data

Underdown, R. B.
Ground studies for pilots.
Includes various editions of each volume.
Includes indexes.
Contents: v. 1. Radio aids — v. 2.
Plotting and flight planning — v. 3.
Navigation general and instruments. —
v. 4. Meteorology.
1. Navigation (Aeronautics) 2. Aids to air navigation. I. Title.
TL586.U48 1993 629.132′51 92–43150
ISBN 0–632–03601–X (v. 1)
ISBN 0–632–03602–8 (v. 2)
ISBN 0–632–03434–3 (v. 3)
ISBN 0–632–03751–2 (v. 4)

For further information on Blackwell Science, visit our website:
www.blackwell-science.com

CONTENTS

Preface vi

1 Basic Radio 1
2 Radio Wave Propagation 16
3 Communications 36
4 Ground Direction Finding (VDF) 44
5 Automatic Direction Finding (ADF) and Nondirectional Beacons (NDBs) 48
6 VHF Omnidirectional Radio Range (VOR) and Doppler (DVOR) 72
7 Radio Magnetic Indicator (RMI) 92
8 Instrument and Microwave Landing Systems (ILS and MLS): Fan Markers 102
9 Basic Radar 119
10 Distance Measuring Equipment (DME): Area Navigation (RNAV) 132
11 Secondary Surveillance Radar (SSR) 145
12 Ground Radars 151
13 Airborne Weather Radar (AWR) 156
14 Radio Altimeters 169
15 Ground Proximity Warning System (GPWS) 174
16 Doppler 180
17 Hyperbolic Navigation Principles: Loran-C 197
18 Decca 202
19 Omega 216
20 Satellite-assisted Navigation (GNSS/GPS) 221
21 Traffic Collision Avoidance System (TCAS) 230

Glossary of Abbreviations 236
Answers to Multi-choice Test Questions 241
Index 242

PREFACE

Following the United Kingdom's signature of the Treaty of European Unity paving the way for a common European Market from January 1993, new syllabuses for pilots' licences have been introduced. The European Civil Aviation Conference, through wide consultations within and between its member states, has produced these harmonised syllabuses setting the required standards. To meet the revised standards, the text of *Ground Studies for Pilots*, which is published in four volumes:

Volume 1 *Radio Aids*
Volume 2 *Plotting and Flight Planning*
Volume 3 *Navigation General and Instruments*
Volume 4 *Meteorology*

and their companion volume, *Aviation Law for Pilots*, have been completely revised.

This volume covers the whole range of radio aids from the properties and propagation of radio waves to the equipment available to pilots on the modern aircraft flight deck and its operation. New material is included on the use of satellites for navigation and communications and on crew alerting systems.

Much appreciated assistance by supplying technical data and illustrations for this edition has been given by British Airways, Smiths Instruments, RACAL Avionics Ltd, GEC-Plessey Ltd, Navstar, Rockwell Collins International and H. W. Cole (for datalink and Mode S). Particularly I would like to acknowledge the valuable help provided by my former colleague Max Johnson for his most constructive suggestions on the revisions and for reading the script.

Self-test questions are provided at the end of many chapters. Specimen examination questions for pilots' licence examinations are available by post from the Civil Aviation Authority, Printing and Publication Services, Greville House, 37 Gratton Road, Cheltenham, Glos. GL50 2BN.

This second reprint of the fifth edition of this volume has been updated to include new material on the future use of MLS and GNSS systems for approach and landing.

Roy Underdown
Max Johnson

Chapter 1
BASIC RADIO

The word 'radio' means the radiation of electromagnetic waves conveying information, and detection of such waves. Within this meaning, such applications as telegraphy, telephony, television and a host of navigation aids are all classified as radio. This volume is primarily concerned with the air navigation aids commonly used world-wide.

The existence of electromagnetic waves was suspected long before Heinrich Hertz conducted his famous experiment in 1887 and demonstrated their presence. As early as 1865 James Clerk Maxwell of King's College, London University produced a paper in which he predicted the existence of these waves. Later, in 1886, a year before Hertz' experiment, Professor Hughes, a scientist in London came very close to the discovery.

However, Hertz not only verified Maxwell's prediction but also established the speed of the radio waves and other properties. He showed that they can propagate in a vacuum, and that they are stopped by a metallic screen (the foundation of our present day radar). He calculated wavelengths for various frequencies and determined the relationship between the two.

Propagation of radio waves

If a source of alternating voltage is connected to a wire (i.e. an aerial) an oscillating current will be set up in the wire, the electrons of which move about a mean position. The electric field present in the wire is accompanied by a magnetic field and at a suitable frequency (in relation to the length of the aerial) both fields radiate efficiently outward from the wire in the form of electromagnetic or radio waves. In the earth environment these disturbances travel approximately at the speed of light, that is,

186 000 statute miles per second or
162 000 nautical miles per second or
300 000 000 metres per second or
300 000 kilometres per second.

As the waves are alternating fields, the terminology involved with alternating currents will be looked into first, extending this to radio terminology.

An a.c. voltage in a wire reverses its direction a number of times every second. Consequently, if a graph of the current in the wire is plotted against time, it will be found that it is a sine curve (see Fig. 1.1).

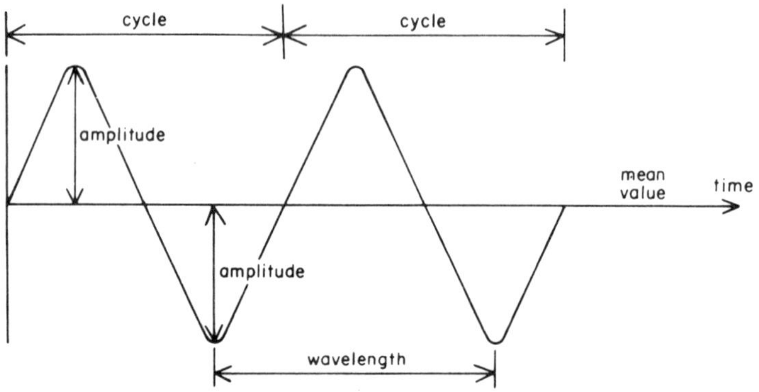

Fig. 1.1 Sine curve and basic terminology.

Cycle. A cycle is one complete series of values, or one complete process.

Hertz. One hertz is one cycle per second. The number of cycles per second is expressed in hertz. (This term is a relatively recent adoption in honour of the above-mentioned eminent scientist.)

Amplitude. Amplitude of a wave is the maximum displacement, or the maximum value it attains from its mean position during a cycle. It is both positive and negative. (That part of the curve in Fig. 1.1. above the mean or time axis is called positive and that part which is below the line is negative.)

Frequency(f). Frequency of an alternating current or a radio wave is the number of cycles occurring in one second, expressed in hertz (Hz). For example, 500 Hz means 500 cycles per second. Since the number of cycles per second of normal radio waves is very high it is usual to refer to them in terms of kilohertz, megahertz and gigahertz as follows:

1 cycle per second	= 1 Hz
1000 Hz	= 1 kHz (kilohertz)
1000 kHz	= 1 MHz (megahertz)
1000 MHz	= 1 GHz (gigahertz)

Wavelength(λ). This is the physical distance travelled by the radio wave during one complete cycle of transmission. It is defined as the distance between successive crests or the distance between two consecutive points at which the moving particles of the medium have the same displacement from the mean value and are moving in the same direction.

Wavelength–frequency relationship

A radio wave travels at a speed of 186 000 statute miles per second, or 162 000 nautical miles per second or 300 000 000 m/sec. The relationship between the frequency and its wavelength is established when it is considered that if a transmission of one hertz is made, the wave will cover a geographical distance of 300 000 000 metres. If two hertz were transmitted (that is, two cycles in one second), two complete cycles will occupy a space of 300 000 000 metres between them. This means that one cycle will occupy 150 000 000 metres, which is also its wavelength. Thus, as frequency is increased, the wavelength is decreased in the same proportion and vice versa; putting this in a formula:

$$\text{wavelength} = \frac{\text{speed of radio waves}}{\text{frequency}} \quad \left(\text{or } \lambda = \frac{c}{f}\right)$$

and

$$\text{frequency} = \frac{\text{speed of radio waves}}{\text{wavelength}} \quad \left(\text{or } f = \frac{c}{\lambda}\right)$$

By use of the above formula it is possible to convert frequency into wavelength and wavelength into frequency. To avoid any errors, at least at the beginning, basic units should be used in the formula. The use of hertz for frequency gives metres for wavelength; the use of metres for wavelength gives hertz for frequency which may then be expressed as kHz or MHz as appropriate for the answer.

Examples

(1) If the wavelength is 1.5 km, what is the frequency?

$$\begin{aligned}\text{Frequency in hertz} &= \frac{\text{speed in m/sec}}{\text{wavelength in m}}\\ &= \frac{300\,000\,000}{1500}\\ &= 200\,000\text{ Hz}\\ &= 200\text{ kHz}\end{aligned}$$

(2) If the transmission frequency is 75 MHz, what is the wavelength?

$$\begin{aligned}\text{Wavelength in metres} &= \frac{\text{speed in m/sec}}{\text{frequency in hertz}}\\ &= \frac{300\,000\,000}{75\,000\,000}\\ &= 4\text{ m}\end{aligned}$$

(3) If the wavelength is 3 cm, what is the frequency?

$$3\text{ cm} = 0.03\text{ m}$$

$$\text{Frequency} = \frac{300\,000\,000}{0.03}$$
$$= 10\,000\,000\,000\,\text{Hz}$$
$$= 10\,000\,\text{MHz or } 10\,\text{GHz}$$

(4) If the frequency is 13 500 MHz, what is the wavelength?

$$\text{Wavelength} = \frac{300\,000\,000}{13\,500\,000\,000}$$
$$= \frac{3}{135}$$
$$= 0.0222\,\text{m}$$
$$= 2.22\,\text{cm}$$

(5) How many wavelengths, to the nearest whole number, of frequency 150 MHz are equivalent to 52 ft?

$$\text{Wavelength} = \frac{300\,000\,000}{150\,000\,000}$$
$$= 2\,\text{m}$$
$$= 2 \times 3.28\,\text{ft}$$
$$= 6.56\,\text{ft}$$

$$\text{The number of times 6.56 ft will go into 52 ft} = \frac{52}{6.56}$$
$$= 8 \text{ (approx)}$$

Now try these.

(1) Wavelength is 3 m, what is the frequency?

Answer: 100 MHz

(2) Express 100 kHz in m.

Answer: 3000 m

(3) Wavelength is 3520 m, what is the frequency?

Answer: 85.23 kHz

(4) Frequency 325 kHz, what is the wavelength?

Answer: 923.08 m

(5) Frequency 117 000 kHz, what is the wavelength?

Answer: 2.56 m

(6) Wavelength 3.41 cm, what is the frequency?

Answer: 8797.6 MHz

(7) How many wavelengths to the nearest whole number is equivalent to 60 ft if the transmission frequency is 100 MHz? (1 m = 3.28 ft)

Answer: 6 wavelengths

(8) If wavelength is 2.739 m, what is the frequency?

Answer: 109.53 MHz

(9) Give the frequency appropriate to a wavelength of 2222 m.

Answer: 135.01 kHz

(10) If the frequency is 1439 kHz what is its wavelength?

Answer: 208.48 m.

Phase and phase difference

Consider a vector, rotating about central axis O and producing an AC waveform. As the vector OR, starting from its position of rest, R, completes one revolution, it will produce one complete cycle of AC. This cycle may be plotted on a horizontal axis, representing 360° (Fig. 1.2).

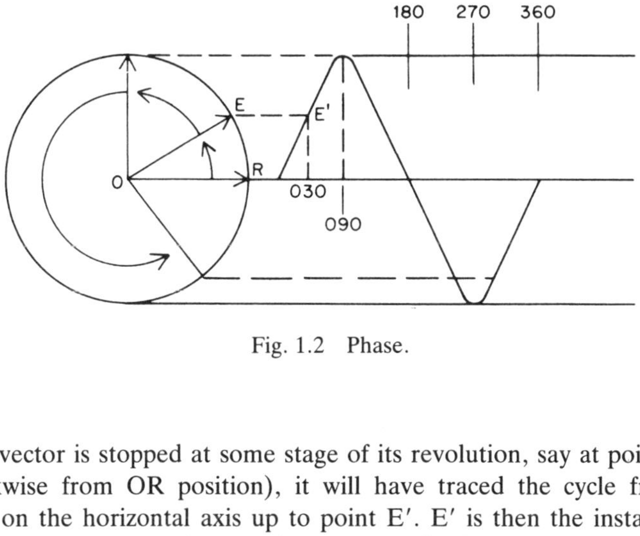

Fig. 1.2 Phase.

If the vector is stopped at some stage of its revolution, say at point E (30° anticlockwise from OR position), it will have traced the cycle from zero position on the horizontal axis up to point E′. E′ is then the instantaneous phase of that cycle. In other words, any stage in the cycle of an alternating current is referred to as its phase.

If two transmissions were taking place on the same frequency, two waveforms would superimpose each other, providing the transmission commenced at the same instant. Then, the two waveforms are said to be in phase. A fractional delay in sending off the second transmission would cause them to be out of phase. To define the term – if two alternating currents of the *same frequency* (therefore, their amplitudes need not be the same) do not reach the same value at the same instant of time, they are out of phase. The phase difference is the angular difference between the corresponding points on

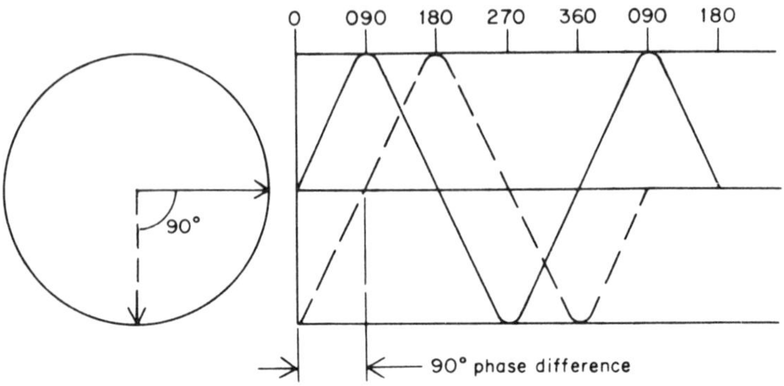

Fig. 1.3(a) 90° phase difference.

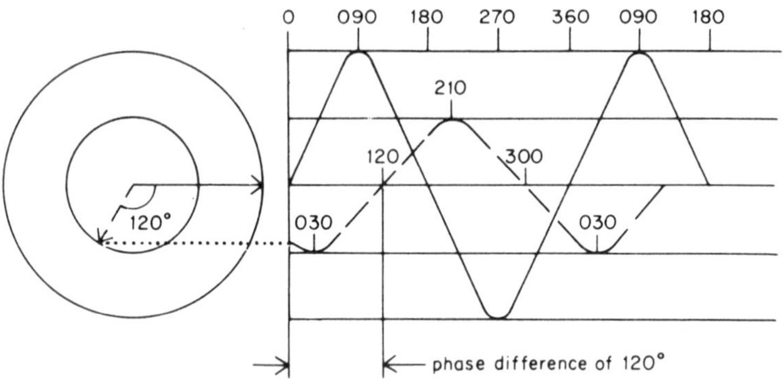

Fig. 1.3(b) 120° phase difference.

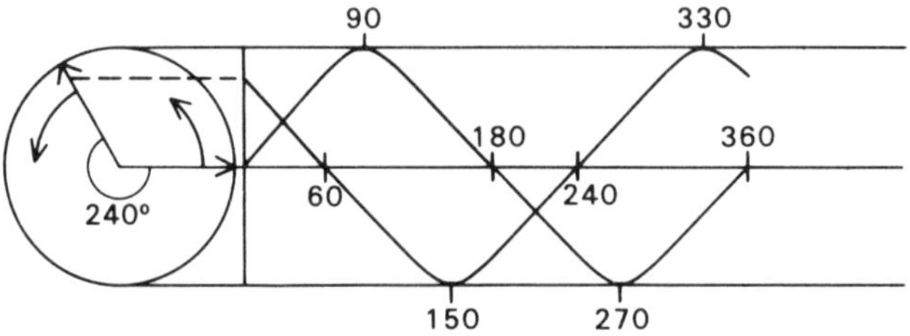

Fig. 1.3(c) 240° phase difference.

the waveform and is measurable. This forms a principle of some of the navigational aids. Two waveforms having any number of degrees of phase difference between them can be drawn by considering revolutions of two vectors placed similarly apart and by tracing their instantaneous values. The point is illustrated in Fig. 1.3(a) and Fig. 1.3(b).

Fig. 1.3(c) is a further illustration showing two signals 240° out of phase.

(With regard to the shapes of the a.c. waves in these diagrams it should be noted that in strict theory, the waves take the shapes of a sine curve.)

Polarisation

As mentioned earlier, when a suitable a.c. is applied to an aerial, electromagnetic waves are radiated from the aerial. These waves alternate with the same frequency as that of the a.c. applied to the aerial. The two components, electric and magnetic, thus radiated travel together at the speed of light. Both travel at right angles to each other (see Fig. 1.4) and also at right angles to the direction of propagation.

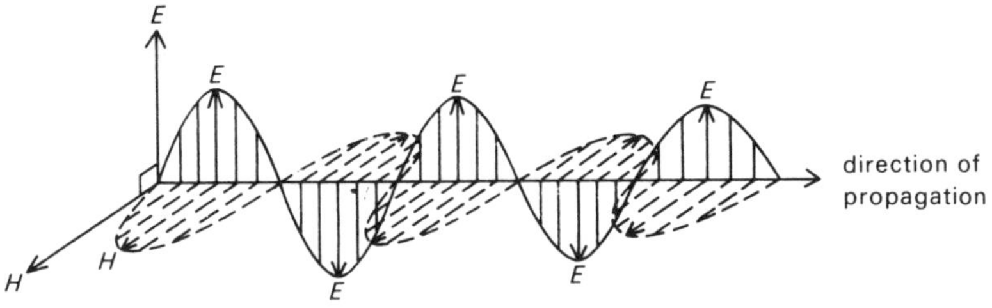

Fig. 1.4 A vertically polarised signal.

When the transmission is being made from a vertical aerial the electrical component, *E*, travels in the vertical plane and its associated magnetic component, *H*, in the horizontal plane and the emission is called vertically polarised. Similarly, for a horizontal aerial the electrical component travels in the horizontal plane, the magnetic component in the vertical plane and the emission is horizontally polarised. Where the electrical and magnetic components spin about the axis of advance, the signal is circularly polarised. This technique is used in reducing rain clutter in radar.

The importance of knowledge of polarisation lies in the orientation of the receiver aerial. A vertical aerial will efficiently receive the electrical component of a vertically polarised signal. If the receiver aerial, on the other hand, was perfectly horizontal it would receive no electrical component. Similarly a horizontal aerial will efficiently receive a horizontally polarised signal.

The vector lengths in Fig. 1.4 represent the field intensity of the signal at a given instant. As the signal travels further the energy spreads out in an ever-increasing volume of space. This is one form of attenuation of the signals – attenuation due to spreadout. The reduction in field strength is governed by the inverse square law in experimental conditions in vacuum. Thus, if the field strength of a point at a given distance from the transmitter measures, say 80 microvolts, then the reading at another point twice the distance from the transmitter will be a quarter of the value, that is, 20 microvolts.

Polar diagram

A polar diagram can be drawn up for any aerial or aerial system to represent the relative values of either field strength or power radiated at various points in both horizontal and vertical planes. For example, the polar diagram of a simple vertical aerial, radiating equally in all directions, is a circle. The polar diagram of two such aerials placed half a wavelength apart would be a figure of eight. Directional aerials radiate most strongly in the required direction and consequently, in any other direction the energy transmitted will be less than the maximum value (see Fig. 1.5).

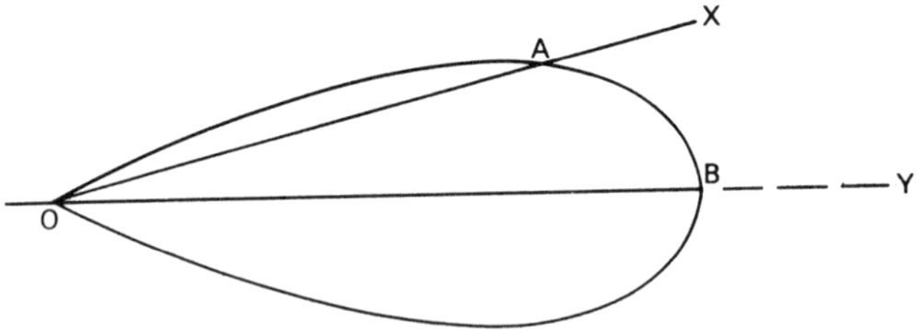

Fig. 1.5 Example of a polar diagram.

The strength of transmission in the predetermined direction is shown by vector OB. On the same scale, vector OA represents the transmission strength in the direction of X.

The polar diagram of a receiver (Rx) aerial similarly gives indication of reception from various directions. An aerial with a circular polar diagram would receive signals equally from all directions whereas an aerial with a polar diagram of a figure of eight (ADF loop aerial) would receive maximum signals from one direction and no signals if the aerial were turned 90° from the original position. In Fig. 1.5, if the receiving aerial was at O, it would receive maximum signals from a transmitter (Tx) in the direction of Y and reduced signals from direction X.

Modulation

A page left blank in a newspaper conveys no information. To convey information it must be impressed with print. A plain radio wave may be likened to the blank newsprint. It can neither be heard nor can it convey information. If it is made audible by use of special components (which will be considered in a later chapter) the only signal heard is a constant audio tone but still nothing is 'read'. Therefore, some form of intelligence must be impressed upon such a wave if it is to convey information. The process of impressing such intelligence is called modulation. It is done in a variety of ways, but since in all cases the radio waves simply act as a vehicle for the information, they are commonly called carrier waves (CW). The waveform of information which is being impressed on this carrier is called a modulating wave. Some of the ways in which the carrier wave may be changed to transmit information are given below.

Keying. This is radio telegraphy. It consists of starting and stopping the continuous carrier so as to break it up into the form of dots and dashes. The communication is by a code, groups of dots and dashes having been assigned particular meanings. The technique is primarily used for long-distance communication; a radio navigation facility may break its carrier to identify itself by dots and dashes. The receiver requires beat frequency oscillator (BFO) facility to make the signals audible.

Amplitude modulation. This method may be used in one of two ways: to transmit coded messages at audio frequencies (AF) or to radiate speech, music, etc.

As the name suggests, in this method, the amplitude of the carrier is varied in conformation with the amplitude of the audio modulating signal, keeping the carrier's frequency constant. To transmit coded information, e.g. identity of a navigation facility, breaks must be caused in the audio. This is done either by keying on/off just the audio tone, or both audio and the carrier.

In Fig. 1.6, audio signal B is impressed on radio frequency (RF) A. Suppose the amplitude of both A and B is one unit. It will be noted that the

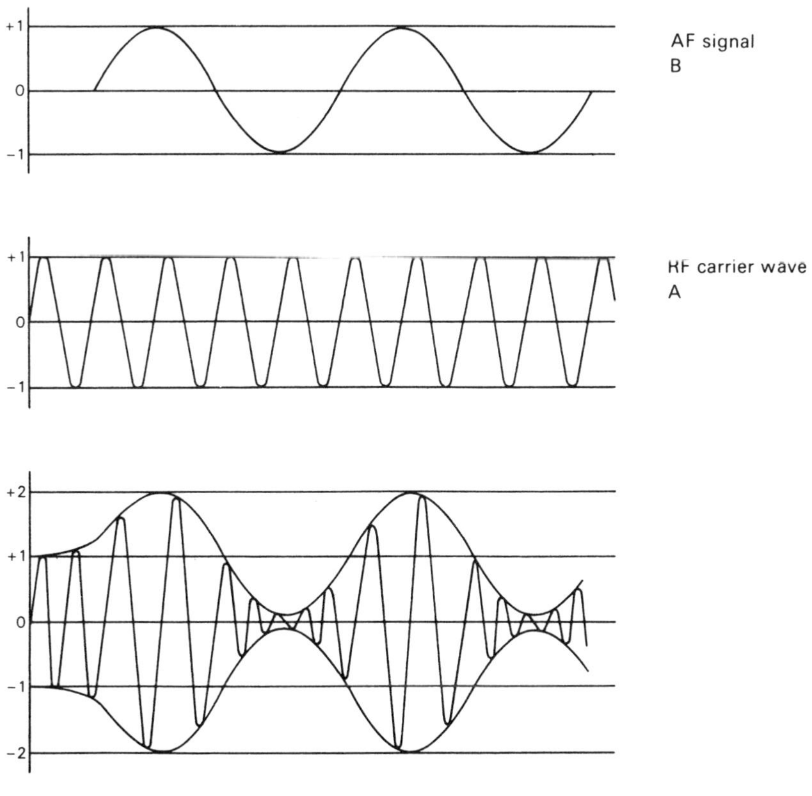

Fig. 1.6 100% modulated RF signal.

resultant envelope of the carrier wave is the picture of the modulating audio wave and that its amplitude has increased to 2 units, now varying between values 0 and 2.

When a signal is amplitude-modulated, its resultant amplitude varies between the sum and difference of the amplitudes of the two waves. In Fig. 1.6 the sum of the amplitudes of A and B is 2 and the difference is 0 and the amplitude of the audio being carried varies between values 2 and 0. This is a measure of the modulation depth. Modulation depth is the extent to which the carrier is modulated and is expressed as a percentage. It is the ratio

$$\frac{\text{amplitude of B}}{\text{amplitude of A}} \times 100$$

In Fig. 1.6 the modulation depth is 100%. If the carrier's amplitude was 2 units and the audio's 1 unit, the resultant audio would vary between 3 and 1 and the modulation depth would be 50%.

The degree of modulation is an important design consideration. Here we are concerned with two factors: the strength of the outgoing audio and the power required to produce it. The variation in the amplitude of the outgoing modulated signal controls the strength of the audio being carried. Thus, a signal with 100% modulation depth will be stronger compared to a 50% modulated signal. High modulation depth would appeal to broadcasters whose speech and music would be heard loudest when 100% modulated. In practice they keep their modulation depth to slightly below 100%. Over-modulation causes distortion in the reception.

As for the power considerations, extra power must be supplied to amplitude-modulate a carrier. The power requirement increases by half for a 100% modulated signal but it falls rapidly when the modulation depth is decreased. Thus, for a given power output and the other conditions being equal, an unmodulated signal will travel further than an amplitude-modulated signal.

Broadcasters in low frequency (LF) and medium frequency (MF) bands employ amplitude modulation, so does civil aviation in very high frequency radiotelephony (VHF RTF).

Frequency modulation. This technique of conveying information was developed in the USA after the shortcomings of amplitude modulation (AM) transmission due to external unwanted noise became apparent during the First World War (1914–1918). It is achieved by varying the frequency of the carrier in accordance with the change in the amplitude of the audio, keeping the amplitude of the carrier constant (Fig. 1.7). The extent of frequency deviation depends on the modulating audio; it is more than the mean carrier frequency when the audio amplitude is positive, and less than the mean when it is negative. The maximum deviation occurs at the positive and negative peaks. In the receiver a frequency discriminator unit detects these deviations and converts them into useful information.

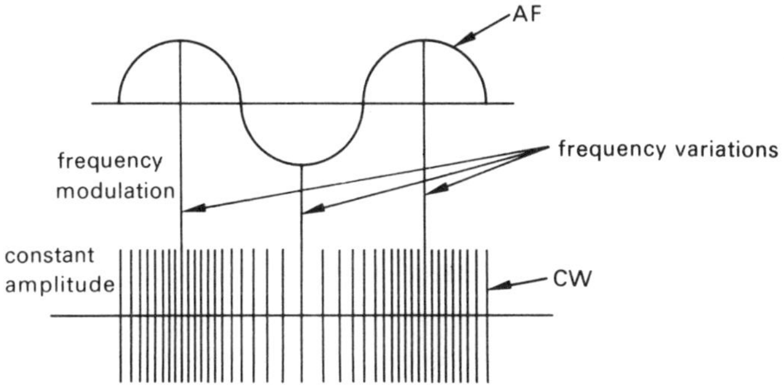

Fig. 1.7 Frequency modulation.

Comparing the technique of frequency modulation with amplitude modulation, frequency modulation (FM) transmitters are simpler than AM transmitters, the necessary modulating power is relatively lower and the reception is practically static-free. This last benefit is due to the fact that the VHF band is practically free from static, and where it is present, it is normally an amplitude-oriented disturbance which enters freely into a vertical receiving aerial. Of the disadvantages, FM receivers are more complex and the modulated transmission calls for a much wider frequency band to cover its multi-sidebands (see below). This is why FM broadcasters operate in the VHF band: the congestion in lower frequency bands would not permit accommodation of the necessary bandwidth. Being in the VHF band, as a side benefit they can cover a complete range of human audio frequencies (up to 15 kHz) and thus provide high fidelity reception whereas in the MF band they would have to be content with staying inside the limit of a spread of 10 kHz.

In civil aviation this technique is employed in radio altimeters which measure height above the surface, and VOR transmits a frequency-modulated carrier. A CW Doppler may find its height using this technique.

Pulse modulation. Pulse modulation is used in radar, and there is a variety of forms of pulse modulations in current use. The modulating pulses in the simplest form amplitude-modulate the carrier, giving it the shape of the pulses.

Sidebands

Sidebands are additional frequencies which occur whenever a carrier is modulated by a frequency lower than itself, particularly audio frequencies.

When a carrier wave is amplitude-modulated, the resultant radiation consists of three frequencies made up as follows:

carrier frequency
carrier frequency + audio frequency
carrier frequency − audio frequency.

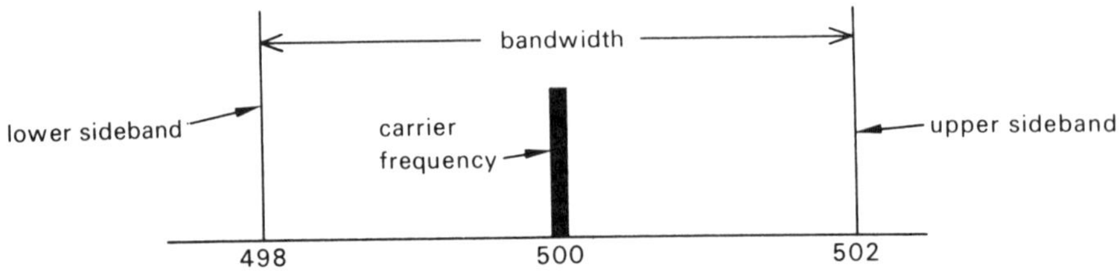

Fig. 1.8 Production of sidebands.

All these frequencies travel together and the new frequencies are called sidebands. In Fig. 1.8, a carrier frequency of 500 kHz is shown being amplitude-modulated by an audio tone of 2 kHz. The resultant side frequencies are 498 kHz and 502 kHz. The former is called lower sideband and the latter is called upper sideband. The complete range, from 498 kHz to 502 kHz is called bandwidth, which is 4 kHz in this illustration.

Unlike AM, a frequency-modulated signal carries with it a multiple of sidebands and consequently its bandwidth is greater.

In the process of modulation, it is the sidebands and not the carrier which carry the intelligence. Therefore, the receiver must be capable of admitting an adequate range of frequencies on either side of the carrier when the carrier frequency is being tuned in. The receiver's bandwidth may be broader than necessary for a particular reception. In this case, by means of a bandpass control this may be narrowed down to reduce external noise or interference from another station.

And because of the sidebands associated with carrier frequency, two stations operating on the same or similar frequencies must have sufficient geographical separation between them to prevent an overlap. This is the primary cause of congestion in the MF and LF bands. The precious frequency space may be utilised more economically by radiating just the single sidebanded (SSB) transmission with an additional advantage of economy in the power requirement, or even both sidebands but each carrying different information, and utilising a common carrier. The following are examples of bandwidth requirements:

speech transmission – 3 kHz
music – between 10 and 15 kHz
radar – 3 to 10 MHz.

Designation and classification of emissions

All radio transmissions used in civil aviation are designated by ICAO according to their description and required bandwidth.

Emission	*Emission designators*
NDB	N0N A1A
	N0N A2A
HF (Communication)	J3E

VHF (Communication)	A3E
VDF	A3E
ILS	A8W
VOR	A9W
DME	P0N

First symbol. Type of modulation of the main carrier. This includes:

N Emission of an unmodulated carrier

and, for emissions in which the main carrier is amplitude-modulated (including cases where sub-carriers are angle-modulated):

A Double sideband
H Single sideband
J Single sideband, suppressed carrier

and, for emissions in which the main carrier is angle-modulated:

F Frequency modulation
G Phase modulation

together with, for emission of pulses:

P Unmodulated sequence of pulses
K Sequence of pulses modulated in amplitude.

Second symbol. Nature of signal(s) modulating the main carrier:

0 No modulating symbol
1 Single channel containing quantised or digital information without the use of a modulating sub-carrier
2 Single channel containing quantised or digital information with the use of a modulating sub-carrier
3 Single channel containing analogue information
7 Two or more channels containing quantised or digital information
8 Two or more channels containing analogue information
9 Composite system comprising 1, 2 or 7 above, with 3 or 8 above
X Cases not otherwise covered.

Third symbol. Type of information to be transmitted:

N No information transmitted
A Telegraphy – for aural reception
B Telegraphy – for automatic reception
C Facsimile
D Data transmission, telemetry, telecommand
E Telephony (including sound broadcasting)

F Television (video)
W Combination of the above
X Cases not otherwise covered.

Information in this context does not include information of a constant, unvarying nature such as provided by standard frequency emissions, continuous wave and pulse radars, etc.

Test questions

(1) Describe an A3E emission and give one radio facility which you associate with it.

(2) Show by means of a diagram two radio signals of the same frequency and wavelength but one 330° out of phase and twice the amplitude of the other.

(3) In what plane does the magnetic field of a radio wave lie if it is:
(a) vertically polarised? (b) horizontally polarised?

(4) By means of suitable diagrams show the following radio emissions:
(a) a frequency-modulated wave
(b) an amplitude-modulated wave.

(5) What do you understand by the terms (indicating type of emission) N0N A1A, P0N, N0N A2A? Suggest one facility to which each might refer.

(6) Show by means of a diagram a radio wave which has N0N A1A emission.

(7) What is a J3E emission?

(8) What do you understand by the term frequency modulation? State one facility which might use this type of emission.

(9) What do you understand by 'sideband'?

(10) Explain briefly the terms phase and phase difference.

(11) A Hertz is:
(a) the frequency in cycles per second
(b) a frequency of one cycle per second
(c) the wavelength corresponding to 1 cycle per second.

(12) If wavelength is 8 mm, the radio frequency is:
(a) 37.5 GHz (b) 375 GHz (c) 3750 GHz.

(13) For a frequency of 200 kHz, the wavelength is:
(a) 1500 m (b) 150 m (c) 1500 km.

(14) In the diagram  time

the two radio waves represented are out of phase by:
(a) 45° (b) 180° (c) 90°.

(15) AM at frequency f_m carried on a transmitted frequency f_c produces:
(a) a sideband of transmission at $f_c + 2f_m$
(b) two sidebands of transmission at $(f_c + f_m)$ and $(f_c - f_m)$
(c) a sideband of transmission at $f_c - 2f_m$.

Chapter 2
RADIO WAVE PROPAGATION

Simple transmitter

The basic components of a simple radio transmitter are shown in Fig. 2.1.

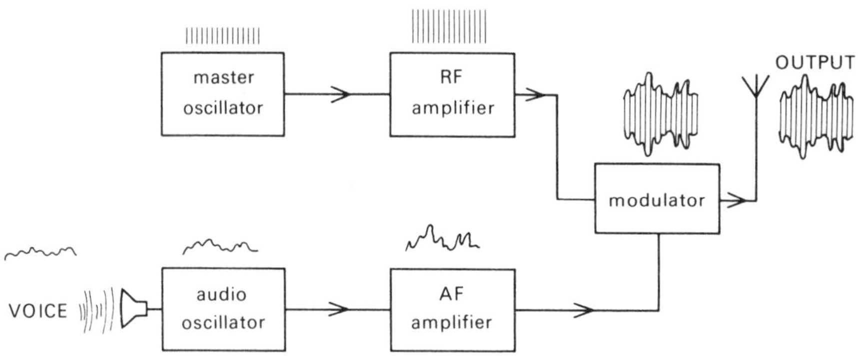

Fig. 2.1 Simple transmitter.

Oscillator. The purpose of an oscillator is to provide a radio carrier wave. At very high frequencies a unit called a magnetron may be used to produce the oscillations.

RF amplifier. The signals produced by the oscillator are too weak for transmission and they must be amplified. This amplification is done at the RF amplifier which is coupled to the oscillator, and the outgoing amplified signals are fed to the modulator.

Microphone and AF amplifier. Similarly, a microphone produces weak audio signals which are amplified by the AF amplifier unit. The amplified signals are then fed to the modulator.

Modulator. In this unit the audio signals modulate the carrier waves by varying the amplitude (amplitude modulation) or the frequency (frequency modulation); the resultant modulated signals are fed for further amplification to the power amplifier.

Power amplifier. Modulated signals arriving at this unit (not shown in Fig. 2.1) are finally amplified (by stages if necessary) to the transmission level.

Aerial. Modulated and amplified signals are fed to the aerial by the power amplifier and the electromagnetic radiation takes place.

General properties of radio waves

(1) In a given medium, radio waves travel at a constant speed.
(2) When passing from one medium to another of different refractive index, the velocity of the waves changes. The waves are also deflected towards the medium of higher refractive index, that is, they change their direction.
(3) Radio waves are reflected by objects commensurate with their wavelengths.
(4) Uninfluenced, radio waves travel in a straight line.

Radio spectrum

The electromagnetic spectrum starts at the lower end of the radio frequencies, that is 30 Hz, and stretches to over ten million million gigahertz where the radiation takes the form of gamma radiation. In this vast spectrum, radio frequencies occupy only a very small part. Different frequencies are found to have different characteristics and in order to identify frequencies having similar characteristics the full range of the radio spectrum is divided into various groups called frequency bands. The frequency bands shown in Table 2.1 are internationally recognised.

Table 2.1

Frequency band	Abbreviation	Frequencies	Wavelength
extremely low frequency	ELF	30–300 Hz	10 000–1000 km
voice frequency	VF	300–3000 Hz	1000–100 km
very low frequency	VLF	3–30 kHz	100–10 km
low frequency	LF	30–300 kHz	10 000–1000 m
medium frequency	MF	300–3000 kHz	1000–100 m
high frequency	HF	3–30 MHz	100–10 m
very high frequency	VHF	30–300 MHz	10–1 m
ultra high frequency	UHF	300–3000 MHz	100–10 cm
super high frequency	SHF	3000–30 000 MHz	10–1 cm
extremely high frequency	EHF	30 000–300 000 MHz	1–0.1 cm
Radar L band		1000–2000 MHz	
Radar S band		2000–4000 MHz	
Radar C band		4000–8000 MHz	
Radar X band		8000–12 500 MHz	

It will, however, be appreciated that these divisions are not 'watertight' divisions and the characteristics of a particular band may overlap above and below the demarcation frequency limit in the table.

The earth and its surround

Before we set out to discuss the type of propagation, the properties and the ranges available in the above frequency bands, let us take a quick look at the physical elements present on and around the earth.

First of all, the shape of the earth: it is approximately a sphere. This means that the horizon curves away with distance from the transmission point, and if the radio waves travelled only in straight lines (as they would, by their basic property) the reception ranges would be limited to 'optical' distance only. This distance is given by the formula $D = 1.05\sqrt{H}$, where D is the range in nautical miles and H is the height in ft. Fortunately, we will soon see that radio waves do curve to a greater or lesser extent with the surface of the earth and in the atmosphere, which means that the above formula is seldom used.

The conductivity of the earth's surface itself varies: sea water provides a medium of high conductivity whereas the conductivity of the land surface depends on its composition. It is fairly high where the soil is rich loam, and very poor in the sands of a desert or the polar ice-caps. The terrain itself varies from flat plains to tall mountains, from deserts to dense jungles.

Surrounding the earth, our atmosphere is rich in water vapour right up to the height of the tropopause. Water vapour is the major cause of the weather and the weather means precipitation, thunderstorms, lightning and so forth. Electrical activity may be expected in any of these attributes of the weather. The other characteristics of the atmosphere, pressure, density, temperature, all vary continually, both horizontally along the surface and with height.

And finally, well above the earth's surface we have electrically conducting belts of ionised layers caused by the ultraviolet rays of the sun.

Radio waves travel best in the free space. On and around the surface of the earth they are influenced to a varying degree by the factors discussed in the preceding paragraphs. We will now study these influences in detail.

Propagation: surface waves

When electromagnetic waves are radiated from an omnidirectional aerial, some of the energy will travel along the surface of the earth. These waves, gliding along the surface are called surface waves or ground waves. As we learnt earlier, it is the nature of radio waves to travel in a straight line. However, in appropriate conditions they tend to follow the earth's surface giving us increased ranges. But, what causes them to curve with the surface?

Primarily there are two factors. One, the phenomenon of diffraction and scattering causes the radio waves to bend and go over and around any

Fig. 2.2 Ground waves.

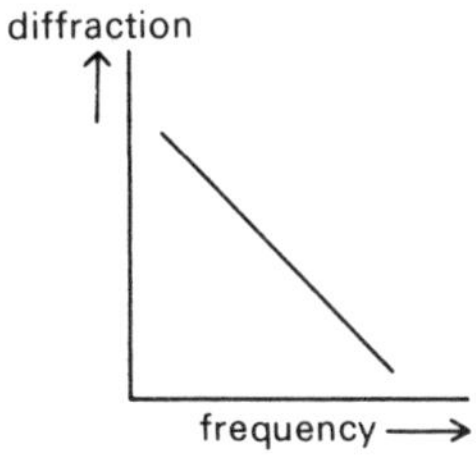

Fig. 2.3 Frequency/diffraction relationship.

obstacles in their path (see Fig. 2.2). As the earth's surface is full of large and small obstacles, the wave-form is assisted almost continually to curve round the surface. The extent of diffraction depends on the radio wave's frequency (see Fig. 2.3). The diffraction is maximum at the lowest end of the spectrum and it decreases as the frequency is increased. At centimetric wavelengths (SHF) an upstanding obstacle stops the wave-front, causing a shadow behind it. It is because of this effect that LF broadcasts give good field strength behind a range of hills but there is no reception on your car radio when going under a railway bridge.

This bending downward is further assisted (the other factor) by the fact that as a part of the wave-form comes in contact with the surface it induces currents in it, thereby losing some of its energy and slowing down. This is called surface attenuation. This slowing down of the bottom gives the wave-form a forward and downward tilt encouraging it to follow the earth's curvature (see Fig. 2.4).

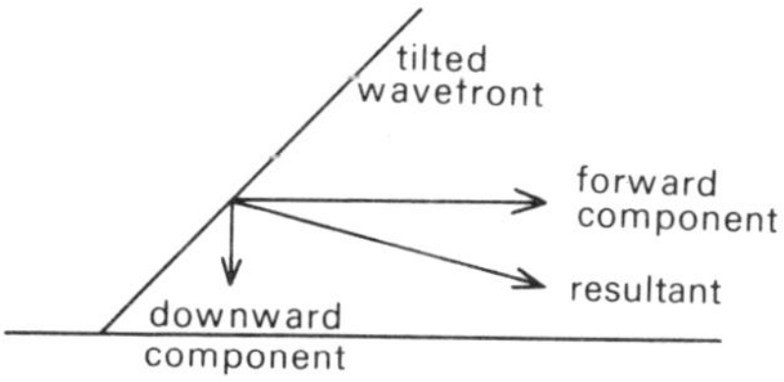

Fig. 2.4 Surface attenuation.

Thus, bending due to diffraction and tilting due to attenuation (imperfect conductivity of the surface) cause the waves to curve with the surface. Waves continue until they are finally attenuated, that is, become undetectable.

Attenuation, in its turn, depends on three factors:

(1) *The type of the surface.* As mentioned earlier, different surfaces have different conductivities. For a given transmission power a radio wave will travel a longer distance over the sea than over dry soil. For example an MF transmitter's range over the sea is nearly double that over the land.

(2) *Frequency in use.* The higher the frequency, the greater the attenuation (see Fig. 2.5).

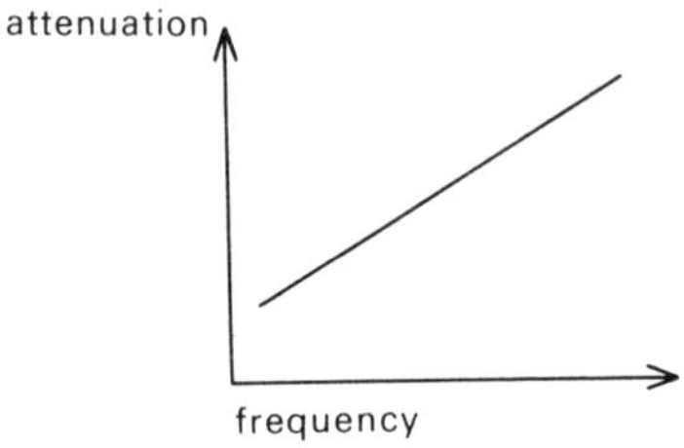

Fig. 2.5 Frequency/attenuation relationship.

(3) *Polarisation of radio waves.* Vertically polarised waves are normally used with minimum attenuation.

In combating attenuation, we have no control over the surface over which the propagation is to be made. The primary consideration therefore, is the choice of frequency. We are now ready to summarise the ground ranges expected from frequencies in various frequency bands.

VLF. Attenuation is least, maximum bending is due to diffraction. Given sufficient power, ranges of several thousand miles may be obtained.

LF. Attenuation is less and the signals will bend with the earth's surface; ranges to a distance of 1500 nm may be expected.

MF. Attenuation is now increasing, signals still bend with the surface and the ranges are approximately 300 to 500 nm, maximum is 1000 nm over the sea.

HF. Severe attenuation, bending is least. The maximum range obtainable is around 70 to 100 nm.

VHF and above. The signals do not bend and the radio waves travel in a straight line, giving line-of-sight ranges.

Disadvantages at low frequencies. Although low frequencies produce very long ranges there are considerable drawbacks which prohibit their inconsiderate employment.

(1) *Low efficiency aerials.* Ideally the length of the transmitter and receiver aerials should each be equal to the wavelength. An aerial approximately half the size of the wavelength is also considered to be

suitable for satisfactory operation. Any further reduction in the aerial size would result in a loss of efficiency. The largest aerials are found in the lowest frequency band – VLF.

(2) *Static* is severe at lower frequencies and additional power must be supplied to combat its effect. The effect of static decreases as the frequency is increased: VHF is considered to be practically free from static.

(3) *Installation and power.* The cost of initial installation is high and subsequent power requirement to maintain the desired range giving satisfactory reception is very large.

It should be noted that the range of a surface wave varies as the square root of its power which may be written in the form of the equation:

$$\text{Range (nm)} = 3 \times \sqrt{\text{Power (watts)}}$$

Sky waves

We have seen how surface waves may be transmitted to varying distances in VLF to HF bands. In these bands, signals may also be received having first been reflected from a huge reflecting layer surrounding the earth known as the ionosphere. These reflected signals are referred to as sky waves and they form the principal mechanism for long range communication.

The ionosphere

The ionosphere is an electrically conducting sphere, completely surrounding the earth. The ultra-violet rays from the sun impinging upon the upper atmosphere cause electrons to be emitted from gas molecules. These free electrons are believed to form a reflecting layer (positive ions would be too heavy to influence electromagnetic waves). Because the absorption of the solar radiation is uneven at different levels in the upper atmosphere, several distinct and separate layers, rather than one continuous zone, are formed. They are given code names D, E and F (see Fig. 2.6). During the period

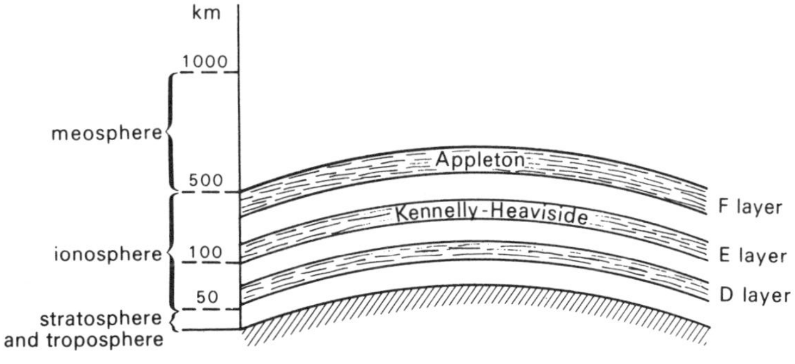

Fig. 2.6 Upper atmosphere.

1901–1930, the E layer was more commonly known as the Kennelly-Heaviside layer, named after its discoverers. The presence of the F layer was established simultaneously by E. V. Appleton in England and A. F. Barnett in the USA and direct measurements were made in 1925, when the name F layer was coined. At present, these belts may be identified either by the code letter or the layer names. Average heights of these layers are as follows, and there are diurnal and seasonal variations.

D layer: 50–100 km, average 75 km
E layer: 100–150 km, average 125 km
F layer: 150–350 km, average 225 km.

The maximum daylight density of the E region is around 10^5 free electrons/cm^3. During the day time the F layer may exist in two separate regions, when the layers are called F_1 and F_2. The maximum value of the electron density in the F layer is around $10^6/cm^3$. When they are separate, F_2 is more persistent than F_1. The electron density in the D, E and F layers varies with time of the day, season of the year and geographical location, as explained below. The overall variation in the E layer is relatively small, but abnormal, sporadic fluctuations may occur all year round, but are more pronounced in summer months. Fluctuations in the F layer are relatively large and irregular, more so during magnetic storms, sun-spot activities and flares. Ionisation in these layers causes refraction, reflection and attenuation of the radio waves, which we will now discuss.

Refraction

Since atmosphere and ionosphere constitute different media, a signal travelling upward will be refracted. If the conditions are suitable it may bend sufficiently to return to the earth. By the accepted usage, the process is called ionospheric refraction. It should be noted that the waves may be refracted by the lower layer and then further refracted by an upper layer. The wave may then return or escape.

Density

The degree of refraction varies directly as the intensity of the ionisation.

The higher the altitude, the thinner the atmosphere. At a height of 60 km the atmospheric pressure is about 0.35 millibars. At such altitudes the solar radiation has a greater effect in breaking down the gas molecules than at lower altitudes where the atmosphere is denser.

Electron density in the E layer is higher than in the D layer; the density in the F layer is higher than in the E layer. Variations in the density occur as follows.

Diurnal activity. In the daytime the solar radiation increases ionic density in all layers and the reflective height moves down. As the sun crosses the meridian the maximum density will be reached. At night as the sun goes below the horizon the process of recombination begins. The D layer, being nearest to the earth and having a relatively denser atmosphere, completely disappears. In the E layer the intensity decreases and the reflecting height rises. The F layer similarly decreases in intensity and finds an intermediate level as one single layer (Fig. 2.7). Sunrise and sunset produce unstable conditions as the layers start falling or rising. These are critical periods for the operation of the automatic direction finding equipment (ADF).

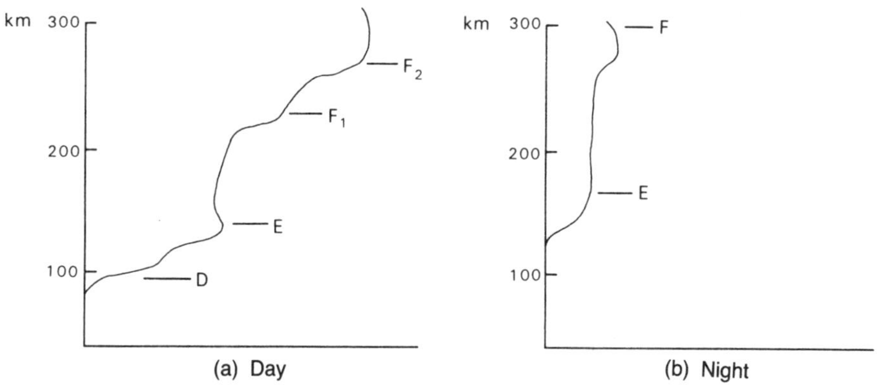

Fig. 2.7 Electron density: N_e.

Seasonal activity. The amount of intensity depends on where the sun is with regard to the position under consideration. There is maximum activity when the sun is closest. Sporadic ionisation occurs in the E layer in summer.

11-year sun-spot cycle. Very marked changes in ionisation occur during this sun-spot activity period. This is due to enhanced ultraviolet and X-radiation from the sun. At this time, ionisation in D and E layers causes an increase in absorption disrupting communication, and signals at VHF frequencies may return.

Attenuation

As mentioned earlier, radio energy is absorbed in the ionosphere. The extent of attenuation depends on various factors.

(1) *Density of the layer.* The greater the density, the greater the attenuation. Maximum attenuation occurs around midday.
(2) *Penetration depth.* The deeper the signal penetrates into the layer, the more loss of energy due to attenuation will occur.
(3) *Frequency in use.* The lower the frequency, the greater the attenuation. This is one of the reasons why a higher frequency is used for communication in the HF band during the day.

Conditions of refraction

Critical angle. The angle at which the signal strikes the layer decides, among other factors, whether the signal will return or not. If it strikes the layer at a small angle to the perpendicular, it will not be refracted sufficiently to return. As this angle of incidence is progressively increased, the signals will bend progressively more until an angle is reached for a given frequency and ionospheric distribution when the first reflection will occur. The angle this wave makes with the normal at the transmission point is called the critical angle (see Fig. 2.8) and the returned wave is called the critical ray or first sky

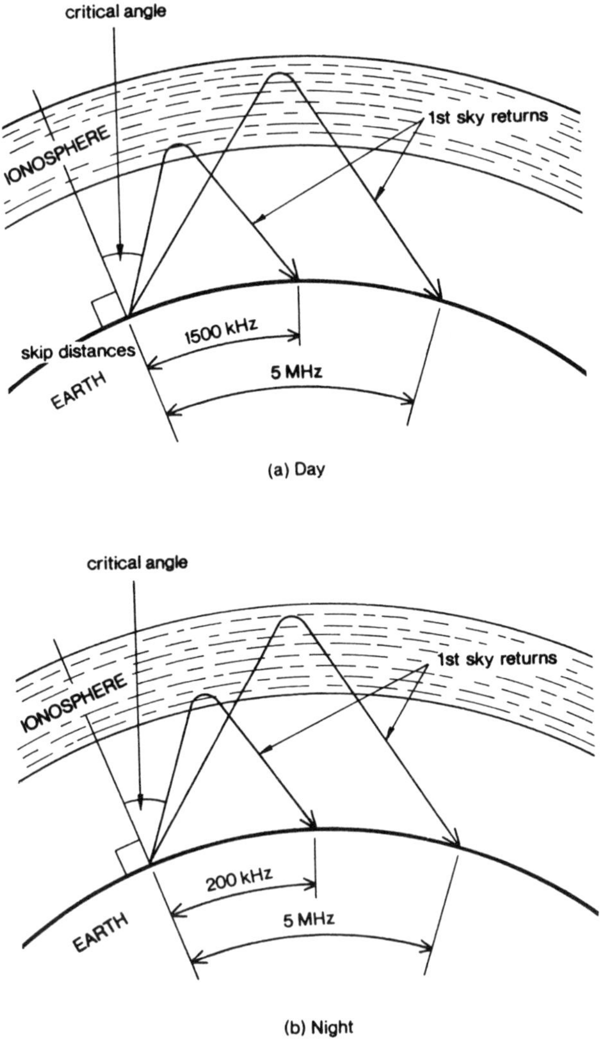

Fig. 2.8 Propagation.

return. At this angle, and higher than this, there will be an uninterrupted flow of sky waves.

Frequency in use. A higher frequency requires a higher electron density to refract it. As ionic density increases with height, higher frequencies will penetrate more deeply into the layer than lower frequencies before returning.

The D layer is not heavily ionised and it will reflect only low frequencies – up to around 500 kHz. For any planned usage of sky waves in this frequency band it should be remembered that the attenuation is predominant.

The E layer is relatively more heavily ionised and will reflect frequencies up to around 2 MHz.

F layer. Frequencies higher than 2 MHz will not be sufficiently refracted in the E layer to return. They will travel to the F layer before returning, thus giving very long ranges. Above 30 MHz, that is VHF and above, the refraction in the layers is insufficient and the signals escape into free space. (The UHF band is used for communication with astronauts in the outer space.) An exception arises in the cases of VHF and UHF during the high solar activity period when the ionisation is extremely dense.

Ranges available

The ranges available from the sky waves depend on the following factors:

(1) *Transmission power*
(2) *Depth of penetration.* This depends on the frequency in use and the ionospheric distribution. The deeper a signal travels before being refracted, the larger the ranges it produces. (Higher frequencies require a higher density to refract them.)
(3) *Critical angle and angle of incidence.* The critical angle will determine the range at which the first sky return occurs, and the prevailing conditions might produce a dead space (see below) where no reception is possible. Similarly, maximum range is given by that wave which leaves the transmitter tangential to the earth.

Figure 2.8 illustrates critical angles and first sky returns at various frequencies during day and night.

Skip distance and dead space

The distance between the transmitter and the point on the surface where the first sky return arrives is called the skip distance. For a given frequency the skip distance varies with the time of day and also the seasons.

Where a signal produces both surface wave and sky wave, there may be an area where no reception is possible. This is because the surface wave's limit has been reached and the sky waves have not started returning (see Fig. 2.9).

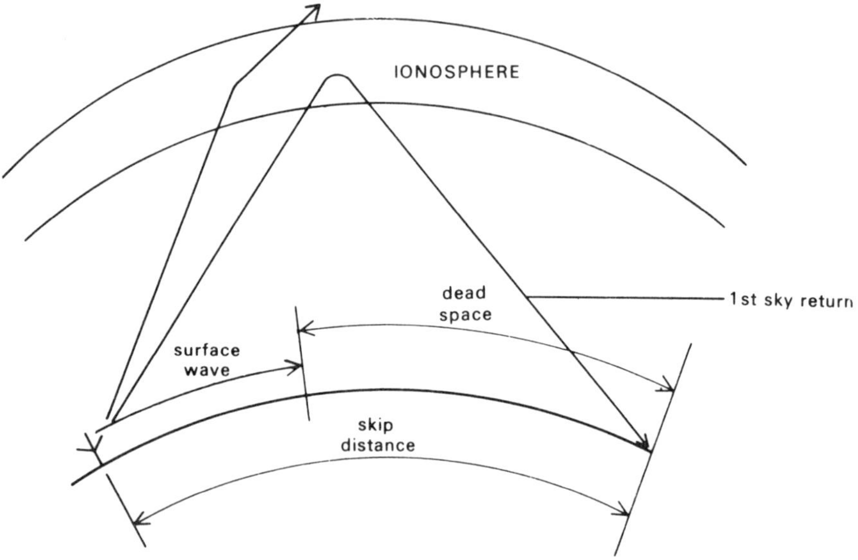

Fig. 2.9 Skip distance.

This area, that is, the area between the limit of the surface wave and the point of reception of the first sky wave is termed dead space.

Dead space is possible in HF where the surface wave is very short and the refraction occurs at higher layers. As the frequency is lowered to MF and LF the surface wave increases whereas the sky wave is returned from lower layers at low critical angles. In these circumstances generally there is no dead space.

Multi-hop refraction

If the returning signals are sufficiently strong they will be reflected from the earth's surface back to the ionosphere where they will be refracted and returned again. This process may continue several times until the signals are finally attenuated by passage through the ionosphere and contact with the earth's surface at the point of reflection (Fig. 2.10). This phenomenon is

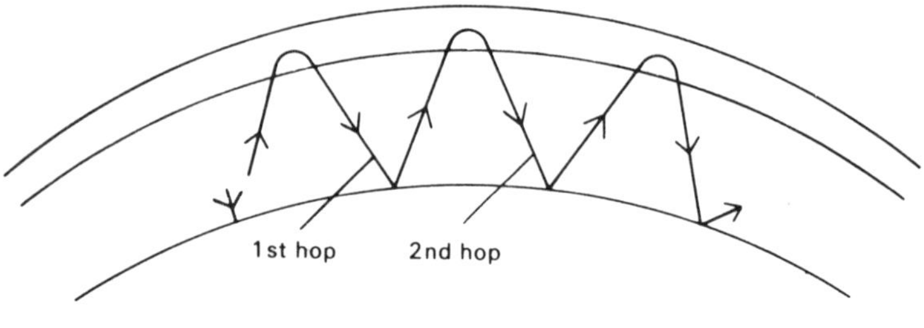

Fig. 2.10 Multi-hop refraction.

known as multi-hop refraction and very long ranges are obtained using this type of propagation. When multi-hop propagation is taking place the first return is called the 1st hop and its subsequent reflections are called 2nd hop and so on. If the angle of incidence is right, the signals can travel round the world and an 'echo' of the previous reception may arrive 1/7th of a second later.

Fading (fluctuation)

Fading is always present, to a greater or lesser extent in sky wave reception because of continuous fluctuations in the ionosphere. The relative phases of the sky waves arriving at a receiver vary in random fashion affecting the amplitude of the output.

It is also possible to receive two sky waves which have travelled different routes or to receive both 1st hop and 2nd hop signals. Since they have not travelled the same distance there will be a phase difference between them. When the two incoming signals are in phase they augment each other giving stronger reception; when they are directly in opposition they cancel themselves out.

Tropospheric scatter

This is the term used when direct refraction from the ionosphere is used, for example in long range military radars, to give increased range.

Summary of properties, using sky waves

VLF, LF. Some sky waves are present during the day and also at night. Low frequencies reflect at relatively low heights, ionospheric attenuation is very large and the propagation is mainly by surface waves.

MF. This frequency band is in mid-position between surface wave and sky wave. Surface wave distances are getting shorter (compared with VLF and LF); the sky waves increase these distances, particularly at night. Sky wave attenuation is less, atmospheric interference is also less but still troublesome. Sky waves in this band are a blessing to such facilities as Loran but are a nuisance to ADF operation.

HF. Surface waves travel only a short distance but very great ranges are achieved using sky waves. Comparative power requirements are less, ionospheric attenuation being only slight.

VHF and above. All frequencies above 30 MHz escape into free space. The ionospheric density is not sufficient to refract them back to the earth.

Space waves

We have seen that above HF neither sky waves nor ground waves may be usefully employed. At frequencies in VHF and above, the only radiated wave which can be used is the one which travels in a direct line from the transmitter aerial to the receiver aerial. This type of transmission is called line-of-sight transmission and it means that if a straight line can be drawn joining the transmitter and receiver, the signals can be received (but see below). The signals thus received are called direct waves.

Sometimes an aircraft may pick up the same signal from two directions: one having travelled direct to the aircraft and the other having first been reflected by the surface (Fig. 2.11). Such a signal is called a ground reflected

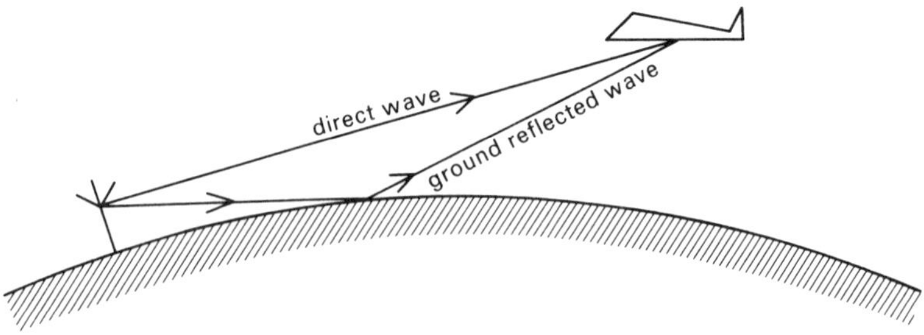

Fig. 2.11 Wave reception.

wave. The direct waves and the ground reflected waves are collectively called space waves. It will be appreciated that when signals are being received from two directions as above, the receiver output will be the vector addition of the two: maximum strength when the two signals are in plase, and fading of signals when the signals are out of phase. The phase relationship between the two signals is governed by the lengths of the two paths and the phase shift at the reflecting point. This phase shift, in its turn, depends on the angle of incidence, polarisation of the incident signal and the conductivity of the reflecting surface.

VHF range

VHF and frequency bands above VHF are straight-line propagation. However, the actual range is slightly better than mere optical range. As we mentioned earlier the distance to the horizon is given in the formula $D = 1.05\sqrt{H}$. The improvement to this range is from the refraction or curving of the waves in atmosphere, due to atmospheric refraction.

The refractive index of the atmosphere, n, is a function of pressure, temperature and humidity. These elements vary significantly in the vertical plane giving rise to diminishing density with increasing height. This means that the refractive index decreases with height. The result is that the radio wave curves towards the regions of heigher density, that is, towards the surface. Thus, signals in VHF and above will be received beyond the optical horizon and the working formula for calculating maximum ranges is:

$$D = 1.25\sqrt{H_T} + 1.25\sqrt{H_R}$$

where D is the range in nautical miles, H_T is the height in feet of the transmitter (amsl) and H_R is the height in feet of the receiver (amsl).

Ground waves

The term 'ground wave' is used to describe all types of propagation except sky waves. Thus, a surface wave is also a ground wave, so is a space wave.

$$\left.\begin{array}{l}\text{direct wave}\\ \text{and}\\ \text{ground reflected wave}\end{array}\right\} = \left.\begin{array}{l}\text{space wave}\\ \text{and}\\ \text{surface wave}\end{array}\right\} = \text{ground wave}$$

Duct propagation or super-refraction

In the foregoing discussion on ranges we saw how a ray curves towards the earth's surface due to the varying refractive index of the atmosphere. If the curvature were slightly greater, the ray would curve sufficiently to travel parallel to the earth's surface and would give very long surface ranges. In normal conditions, however, this is not so, but meteorological conditions can exist from time to time at an altitude which could produce an increased curvature.

An increase of temperature with height sometimes occurs through a few thousand feet which is then known as an inversion. The condition may even start at the earth's surface when it is called a surface inversion. A similar situation can arise from an abnormal distribution of water vapour with increase of height near the surface (the hydrolapse). These conditions affect the relative permittivity (ε) of the air (formerly called its dielectric constant), a measure of its insulation property which is low in cold dry air and high in warm moist air. Normally with increase of height, the air's temperature and humidity content decrease giving a corresponding decrease of permittivity. In turn, this gives greater refraction near to the surface causing a direct radio wave to exceed the tangential horizon. In the abnormal conditions of a temperature/humidity inversion, the permittivity increases with increase of height to the top of the inversion and then suddenly decreases in the colder, drier air above the inversion so resulting in super-refraction from the top of the inversion layer. In the VHF band and above, freak reception ranges of several hundred miles can be caused. The meteorological conditions suitable for such anomalous propagation are:

- warm, dry air blowing over cool sea
- subsidence
- pronounced radiation cooling.

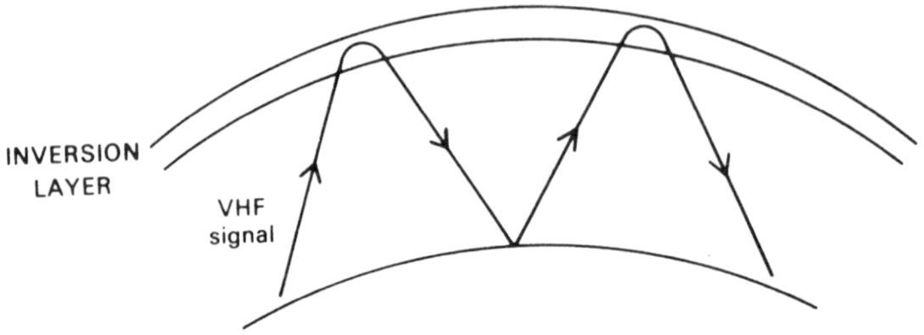

Fig. 2.12 Duct propagation.

This type of propagation is called duct propagation or super-refraction (see Fig. 2.12).

The climatic conditions most conducive to the formation of duct propagation are usually found in tropical and sub-tropical latitudes. This is because the land mass can get extremely hot during the day followed by a rapid cooling at night.

Sporadic-E reception

Patches of intense ionisation occur from time to time in the E layer. The effect occurs all year round but is highly pronounced in summer. The patches are comparable to clouds in the troposphere and drift with time. Sometimes they last for a few minutes, other times for a few hours. Signals up to about 150 MHz are scattered from the 'blobby' sub-surface of the patches and may be received up to 1000 to 1500 nm away. The phenomenon presents an ideal opportunity for long-range communication but for a DF application, due to its scattering effect, the direction measured could be as much as 180° in error.

A scattering effect is also present in the troposphere in regions where air turbulence is prevailing. The eddies are found to contain discontinuous refractive indices, which result in abnormally strong signals on one occasion and equally strong interference on another.

Properties – miscellaneous

Static. Static consists of amplitude-oriented spurious electromagnetic waves ranging in frequencies from a few hundred hertz to several thousand kilohertz. Atmospheric static is present in a cumulonimbus cloud, thunderstorms and all kinds of precipitations. Lightning discharge is a most common source of static.

Static also reaches us from interstellar space varying in frequency between 15 MHz and 100 MHz. It is high in field strength and fairly constant in amplitude. This is the very substance on which radio astronomers thrive.

Radio noise. There are two sources of radio noise: avoidable and unavoidable.

Avoidable noise arises from use of poor quality components, bad contacts and similar causes.

Unavoidable noise occurs both internally and externally. Internally, it is due to small thermal motions in the components. These are always present to a varying degree when the equipment operates at temperatures over absolute zero temperature. They show up as 'grass' on a radar screen or a hissing noise on a loudspeaker. The sight of grass on a radar screen is a source of delight to the operator who is thereby assured that at least the thing works up to that stage.

Externally, the source of noise is the electrical disturbance in the atmosphere and the outer space, in the form of static. Atmospheric static is important up to 30 MHz (end of HF band), extra-terrestrial noise may cause interference up to 100 MHz.

Seasonally, the effect of static is stronger in summer than in winter; geographically, it is stronger in the tropics than in temperate zones; and diurnally, it is stronger at night due to sky waves which travel a great distance picking up static en route.

Radio interference

Interference in reception may be caused by a variety of reasons. Due to congestion of radio frequencies, same and similar frequencies may be allotted to other stations which may cause direct interference. The remedy is that the propagation ranges of such stations are strictly regulated by the various States (protection range, altitude – see later chapters) and they determine the use of the station within its promulgated range altitude.

Stations are also given a fixed tolerance on transmission frequency and they must maintain the transmissions within these limits. Interference will be caused if the transmitter is allowed to drift beyond the limit, and if this comes to your notice the remedy is to report the matter to the authorities.

Man-made interference may exist where there are heavy industries in the vicinity of an aerodrome or a facility transmitter. Harmonics of 50 Hz a.c. can enter into instrument landing system (ILS) localiser radiation (150 Hz modulation), an overhead cable can affect an NDB radiation and even small components such as switches, motors and ignition systems can cause electrical noise resulting in interference. A moving vehicle can affect category 2 ILS directional characteristics. The remedy for this type of interference is to suppress it at the source. This type of interference does not affect high frequencies.

Signals/noise relationship

It will be apparent from the foregoing discussion that there is always some interference/noise present in the reception of radio signals. A receiver must be able to admit the selected frequency together with its sidebands. The bandwidth of a receiver may be too large. On other receivers it may be possible to select the bandwidth by means of a switch on the control box. If the interference is due to another more powerful transmitter on a nearby frequency, it may be cured by excluding interfering signals by narrowing down the reception bandwidth. Similarly, static interference may be minimised, but care must be taken not to exclude the wanted signal. Generally, do not select a bandwidth any higher than necessary.

The ultimate incoming signals (both wanted and unwanted) are expressed as signal/noise ratio. This ratio is given in decibels. A decibel (dB) is a unit which is extensively used in electronics to provide a measure of gain/loss in terms of power or voltage. In terms of voltage (in which the field strength is measured), the relationship is

$$\text{gain/loss in dB} = 20 \log (E_1/E_2)$$

Illustrating this: ICAO recommends that the States provide a protection to the NDB transmissions by ensuring a signal/noise ratio of at least 3 to 1. By means of the above formula we can work out what this means in terms of decibels.

$$\begin{aligned} \text{Gain} &= 20 \times \log 3 \\ &= 20 \times 0.4771 \\ &= 9.542\,\text{dB or } 10\,\text{dB} \end{aligned}$$

Thus, the field strength of the wanted signal is 10 dB higher than the unwanted signal.

Spurious noise may be reduced by such devices as wick dischargers and receiver filters.

Attenuation

Surface. At low frequencies, the earth's surface acts as an absorber. Signals lose their strength as they travel forward. Attenuation also slows down a signal's velocity. The following velocities may be assumed for a 100 kHz signal:

299 750 000 m/sec over a good conducting surface (water),
299 250 000 m/sec over a poor conducting surface (sand).

At metric (VHF) and centimetric (SHF) wavelengths, the ground acts as a reflector.

Atmospheric. Attenuation in the atmosphere occurs in two ways: by direct absorption in gases and by absorption and scattering by solid and liquid particles. Wavelengths of under 10 cm are affected by these elements and the effect progressively increases as the wavelength becomes shorter. Raindrops absorb signals at wavelengths immediately below 10 cm whereas signals of 5 mm wavelength are absorbed by oxygen molecules. This following are the typical losses as a signal travels a distance of one kilometre:

	Wavelength 5 cm	Wavelength 5 mm
Heavy rain	0.7 dB	7 dB
Fog, vis 100 ft	0.4 dB	3 dB

Summary

A summary of the frequency bands, characteristics of propagation and associated facilities is given in Table 2.2.

Table 2.2

Wavelength denomination	Frequency band	Characteristic	Application
myriametric	VLF	*Propagation*: ground and sky wave *Range*: 4000 nm by surface wave *Static*: severe *Aerials*: very large	very long-range navaids Omega BT long-range communication
kilometric	LF	*Propagation*: ground wave by day; ground wave and sky wave by night *Range*: 1500 nm by surface wave *Static*: less, but still troublesome *Aerials*: large	Decca, NDB, Loran, ADF, Broadcast
hectometric	MF	*Propagation*: as LF *Range*: 300–500 nm, max. 1000 nm over water by day; sky wave arrives a little beyond the ground wave, giving greater ranges at night *Static*: present and troublesome	NDB, Broadcast, Loran, ADF
decametric	HF	*Propagation*: mainly by sky wave, day and night *Range*: ground wave approx. 100 nm; sky wave approx. 3000–4000 nm *Static*: still present but less troublesome	long-distance communication RTF
metric	VHF	*Propagation*: space wave, giving line-of-sight ranges *Static*: negligible	VHF RTF, ILS LOC, VDF, VOR, fan markers

Table 2.2 *Continued*

Wavelength denomination	Frequency band	Characteristic	Application
decimetric	UHF	*Propagation*: line of sight *Attenuation*: attenuation from water vapour starts above approx 1000 MHz but not important	ILS GP, DME, surveillance radar, UHF RTF
centimetric (also microwave)	SHF	as above, attenuation increase	PAR, surveillance radar, Doppler, AWR, radio altimeter
millimetric	EHF	as above, attenuation severe; very small aerials	ASR, experimental radar

Test questions

Check these out before going further – you won't regret it!

(1) Explain the following terms:
(a) skip distance (b) ground (or surface) attenuation
(c) duct propagation (d) ionospheric attenuation
(e) atmospheric attenuation (f) tropospheric scatter.

(2) By means of a suitable sketch, show the following:
(a) critical angle (b) skip distance (c) dead space
(d) the wave giving the longest range by one hop.

(3) Explain with the aid of a diagram the meaning of the term 'multi-hop refraction' and show the 1st, 2nd and 3rd hop on the diagram.

(4) Discuss the advantages and disadvantages of sky waves to the present navigation aids.

(5) Give the names and ranges of the frequency bands which lie between 30 kHz and 3000 kHz inclusive, and give one example of a radio or radar facility appropriate to each band.

(6) In which of the bands in question 5 would you expect to find:
(a) the most complete freedom from static interference?
(b) sky waves reliable enough for use either by day or night?
(c) sky waves by night, but some or only weak ones by day?

(7) When utilising sky wave propagation, what is the relationship between frequency and range? What factors affect this relationship?

(8) Give a brief description of the ionosphere. What effect does it have on the reception of radio signals?

(9) State briefly what you understand by atmospheric static.

(10) Comment briefly on the expression 'VHF line-of-sight transmission' used in radio.

(11) An aircraft flying at 9000 ft could expect to be within VHF contact of an airfield 1600 ft amsl at a maximum range of approximately:
(a) 98 nm (b) 138 nm (c) 168 nm.

(12) Static interference is most likely to be severe on:
(a) VHF (b) LF (c) UHF.

(13) A major factor which influences sky wave propagation particularly is:
(a) power (b) terrain (c) time of day.

(14) Radio waves on the SHF band are propagated mainly by:
(a) ground wave (b) direct wave (c) sky wave.

(15) In the ionosphere, the F layer:
(a) is the lowest ionised layer
(b) is at all times more weakly ionised than other layers
(c) may be split into two layers by day.

Chapter 3
COMMUNICATIONS

Long-range communication – choice of frequency band

To achieve communication on the basis of global distances, the choice must lie in the bands between VLF and HF, the frequency bands above HF being 'line-of-sight' propagation.

Starting at the lowest end, we could obtain very long ranges in the VLF and LF bands and settle for one of them without further ado, but there are some inherent disadvantages in the employment of these bands. Just two requirements, of aerial and power alone are sufficiently forbidding to spur the researchers to investigate alternative possibilities.

These possibilities are MF and HF. Of these two, HF is considered to be far superior:

- aerials are shorter and less expensive to instal
- static noise is less than in MF and tolerable
- by using sky waves day and night, very long ranges are obtained for relatively less power
- higher frequencies suffer less attenuation in the ionosphere
- efficiency is further increased by beaming the radiation in the direction of the receiver.

In the early days of radio, experiments were made on the utilisation of long waves for communication purposes, but the benefits of the short waves soon became apparent and by the late 1920s the rush was on for short waves.

HF communication

The principle of HF communication relies on choosing a frequency appropriate for a given set of ionospheric conditions that will produce the first return at the required skip distance from the transmitter. If the height of the refracting layer is known, the ray's path from the transmitter to the receiver via the ionosphere can be plotted and, from this, the angle of incidence the ray makes at the ionosphere can be ascertained. The frequency to use, so that the ray travels along this path, is derived from the knowledge of the angle of incidence, θ, and the critical frequency, f_c. The critical frequency is that frequency which just starts to escape at vertical incidence to the ionosphere. The mathematical relationship between the two, $f_c \times$ secant θ, gives us what is called maximum usable frequency (MUF). This is the highest frequency available for that predetermined distance, prevailing density and height of penetration. If it is increased any further, the signal will escape. If it is lowered considerably, excessive attenuation will cause unacceptable power loss. When this limit is reached, it is called the lowest usable high frequency (LUHF).

In practice, graphs and nomograms are made available to the radio stations from which this value is directly extracted. The graphs take into consideration such factors as the station's position in latitude and longitude, time of the day, season of the year, density of the ionosphere and any abnormal condition prevailing and the distance at which the first sky return is required.

It will be appreciated that because of the diurnal variation in the ionospheric density, if transmission is continued at night on daytime frequency, a wider skip distance will result, leaving the target receiver in the 'dead space'. This is because at night

(1) the reflection height increases, and the wave is returned from a higher level, giving a greater skip distance
(2) the density of the layer decreases which requires the same wave to travel still higher in denser density layers before being returned, giving the same consequence as (1).

For these reasons, the working frequency is lowered at night. This lowering of the frequency adjusts the skip distance because

- lower frequencies reflect from lower levels
- lower frequencies require a smaller critical angle.

It may be pointed out here that in lowering the frequency at night the signal is not being subjected to an increased attenuation since the density is less at night. In practice, the night frequencies are approximately half of the daytime values.

On the matter of the choice of frequency, if you are calling a station without success, but you are hearing another station which is at a greater distance in the same direction, you are operating on a frequency which is too high. The station you wish to contact is in the zone of skip distance. Lower the frequency.

The frequency band allotted to commercial aviation ranges from 2 MHz to 22 MHz – in practical use it is limited to around 18 MHz. The ground stations publish a number of frequencies for use, and the communication is generally addressed to the ATCC or ACC. The service range depends on the requirement (around 1000 nm). The transmission is amplitude-modulated and an SSB emission is used to economise in *power* and *channel space*.

In the early days when MF/HF WT was in the forefront, aircraft were equipped with a trailing aerial. It consisted of a coil of wire which was wound out and held downward by a weight. Normally it disappeared at the first sight of thunder or lightning. In another system a permanently fixed wire was used, stretching the length of the fuselage. These aerials have now been replaced by recessed aerials conveniently located to give an all-round reception.

Because of the expense of the initial installation, use of HF RTF is limited at present to the airlines and other large aircraft. Power-wise, a mere 100-watt transmitter would provide transatlantic communication.

Factors affecting HF range

- Transmission power
- Time of the day; this governs the density and height of refraction
- Season of the year; this has bearing on the density
- Any disturbances in the ionosphere
- Geographical location
- Frequency in use; this determines the critical angle and the penetration depth.

Short-range communication – choice of frequency band

The requirement here is to provide communication at 80 nm range at 5000 ft and 200 nm at 20 000 ft. As these are very short ranges, frequency bands from VLF to HF may be ruled out.

Up the spectrum from VHF, it is best to choose the lowest frequency band from the aerial consideration. The aerial requirement gets more complicated as higher frequency bands are reached. Even in VHF, VOR employs a special aerial whereas VDF is a ground installation. For a simple aerial as used in RTF, the signal strength received at a given range is proportional to the wavelength. Thus, a larger wavelength (i.e. lower frequency) would provide a better field strength.

VHF communication

The VHF band is chosen for RTF communication at short ranges, the operative frequencies being kept at the lower end of the band, i.e., 117.975 MHz to 137.000 MHz. Within this band, 760 communication channels are or will soon be available at 25 kHz separation. The transmission is amplitude-modulated, the type of emission being A3E. A transmitter producing 20 watt power would be considered quite adequate for maximum ranges. In the early 1990s, the former upper limit of 136.000 MHz was extended to 137.000 MHz and in the European region (EUR) it was agreed that to relieve acute channel availability problems, the extra 40 channels thus available would be used as follows:

(1) the first 32 channels (136.025–136.775 MHz inclusive) for national and international ATC purposes
(2) the next four channels (136.800–136.875 MHz inclusive) for international operational control
(3) the top four channels (136.900–136.975 MHz inclusive) reserved for datalink purposes (see chapter 11).

VHF is practically free from static, but being vertically polarised, the signals do pick up some background noise. If absolute clarity of reception is required, the choice should be shifted to UHF where room may be available to accommodate FM sidebands.

Factors affecting VHF range

(1) Transmission power both at aircraft and ground station.
(2) Height of the transmitter.
(3) Height of the receiver.
(4) Obstacles at or near the transmission site will block the signals or scatter them with inevitable attenuation.
(5) Any upstanding obstruction in the line of sight between the aircraft and the ground station will have an effect similar to (4) above.
(6) In certain circumstances the aircraft may receive both direct and ground reflected waves which may cause fading or even short term loss of communication.

Selective calling system (SELCAL)

This system of communication relieves a pilot from the tiresome task of maintaining a continuous listening watch on the RTF while in flight. It is most beneficial when an aircraft is flying in peaceful areas, e.g. on a long ocean crossing, where the only need for the RTF is to make its periodic position reports. The advantage of the facility is taken by installing a SELCAL receiver in the aircraft. The Air Navigation Order prescribes rules with regard to its use.

(1) The ground station is informed that you intend to use SELCAL
(2) the ground station must not raise any objection, and
(3) the particular ground station is notified as capable of transmitting SELCAL codes.

For its use outside of the UK ensure that the ground station concerned is designated as transmitting a signal suitable for the purpose.

ATC must be informed of the codes carried in the aircraft and a preflight functional check must be carried out. If at this stage or at any stage en route it is thought that either the ground or the airborne equipment is unserviceable, listening watch must be resumed.

When on SELCAL, if the ground station wishes to contact you, it will transmit a group of two coded tone pulses. The decoder circuit will accept the signal if it is meant for it, and activate the cockpit call system by flashing a lamp or by ringing a bell or a combination of both.

Secondary surveillence radar (SSR)

Secondary surveillance radar is another method of communication from ATC to the aircraft. With the use of SSR, ATC derives an aircraft's identity, flight level and follows its track. For fuller information on SSR see chapter 11.

Aircraft communications addressing and reporting system (ACARS)

Usually operating in conjunction with an aircraft's flight management system (FMS) (see *Ground Studies for Pilots* Volume 3), ACARS is a communications datalink system between the airline's ground operations base and its aeroplanes. It uses the aircraft's VHF communication system to send and receive information. Usually this will be the No. 3 VHF and dedicated just to the ACARS. However some systems have the No. 3 VHF with an additional VOICE/DATA selector switch. When switched to VOICE the system is used just as any other VHF communication channel. When switched to DATA, the ACARS handles the downlink from the aircraft and the uplink from the ground station to the aircraft.

For the student pilot, air/ground/air messages usually relate just to ATC matters but on a commercial flight with several hundred passengers on board, messages on the 'company channel' are common concerning passenger services, maintenance services, fuel state and so on. With ACARS, the spoken messages which would have had to be made by the crew are replaced by a compressed format which is transmitted in about a second in the downlink. The message can be an automatically compiled report or one prepared by the pilots on the control and display unit (CDU) of the FMS. Very basic ACARS messages for example are the routine Out Off On In (OOOI) times, generated and transmitted:

Out – at the time the aircraft leaves the stand, with doors shut and the parking brake released
Off – at take-off when the main gear moves to the retracted position
On – on touchdown when the aircraft's weight is on the wheels
In – at the terminal unloading bay, parking brake on and a door opened.

Depending upon the company policy some ACARS equipment can automatically transmit routine reports (without involving the flight crew at all) as frequently as the operator needs. Typical routine reports include aircraft and crew identification, flight plan and current flight conditions, engine performance, fuel state, systems performance, maintenance items and passenger services.

When the operator uses the ACARS uplink, it may well be able to obtain the required information from the FMS computer. If it does require an input from the flight crew, the SELCAL link alerts the crew. The pilots then read the message displayed on the FMS CDU and respond via the downlink. Aircraft may have a printer to print out messages to pilots.

With the congestion that occurs on VHF, even transmitting a one-second formatted ACARS message may coincide with another transmission on the dedicated ACARS frequency. A management unit (MU) in the aircraft ACARS will check that the channel is free before starting its transmission, but if two aircraft do transmit simultaneously the messages received will be garbled. If this occurs repeatedly, the FMS will display NO COMM to the crew. Otherwise both downlink and uplink messages are automatically acknowledged within both ACARS airborne and ground units.

Communication via satellite (SATCOM)

Although once it was a novelty we now regularly see on our televisions, programmes beamed to our homes via satellites. Aviation is also a satellite user for communication, primarily through INMARSAT. There is a constellation of geostationary satellites provided by the International Maritime Satellite Organisation for the relaying worldwide of telecommunications for aviation, shipping, and land mobile users. In fact, with the geostationary satellites over the equator, worldwide effectively means in all longitudes, between 80°N and 80°S (Fig. 3.1).

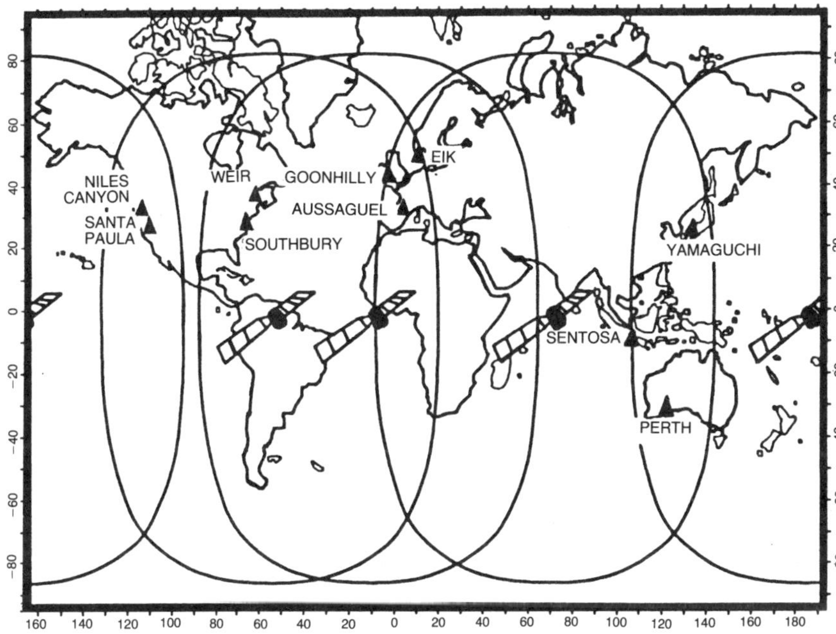

Fig. 3.1 Satcom coverage and ground earth station networks.

To use SATCOM, aircraft have to have special antennae to receive/transmit the communication beam to the appropriate satellite. The transmission is digital in the L-band for which the speech waveform has been digitised at 9600 bits/second, so is virtually unaffected by meteorological conditions or geographical position. As far as flight crew and passengers are concerned, there is no difference from other aeronautical telephone handsets or car-phones.

On the ground, there is a network of ground earth stations (GES) which are positioned so that they service each of the four satellite regions or segments and link into the conventional public and private telephone networks.

Legislation on the use of radio

The provisions (paraphrased) below, are prescribed in the Air Navigation Order:

(1) An aircraft radio may not be operated either on the ground or in flight unless it is licensed and only by a licensed or permitted person and only in accordance with the terms of the licences/permission.
(2) In flight a continuous radio watch must be kept on the radio communications apparatus by a crew member or by SELCAL (subject to the conditions for SELCAL already listed).
(3) Radio or radio-navigation equipment must be operated by flight crew as instructed by ATC or as the notified procedure.
(4) Aircraft radio must not be operated so as to cause interference with communication or navigation services and emissions must not be made except as follows:
    (a) emission class and frequency must be appropriate to the airspace in which the aircraft is flying
    (b) distress, urgency and safety messages
    (c) messages and signals relating to the flight of the aircraft
    (c) such public correspondence messages as are permitted under the aircraft radio station licence.
(5) A telecommunications log book must be kept if radio communications are made by WT.
(6) In a public transport aircraft registered in the UK, the flight crew may not use a hand-held microphone for RTF (or intercom in the aircraft) whilst flying in controlled airspace below FL 150 or while the aircraft is taking off or landing.

Test questions

(1) Give the main factors which affect the range of ground-to-air communication in:
    (a) HF band (b) VHF band.

(2) Explain why day and night frequencies are different in HF.

(3) Radio ducting is of most significance on the frequency bands:
    (a) VLF to MF (b) VHF and above (c) MF and HF.

(4) The MUF between two specified places at a particular time is:
    (a) the frequency which gives the least radio interference
    (b) the maximum frequency which can be used
    (c) the maximum frequency which is reflected by the ionosphere.

(5) Heading 270°T around dawn with a choice of two frequencies of 9 MHz and 5 MHz for HF communications, to contact a station it would be better to use:
    (a) 9 MHz for the station ahead
    (b) 5 MHz for the station behind
    (c) 5 MHz for the station ahead.

(6) For a given HF frequency, skip distance will normally:
    (a) have no diurnal variation

(b) be greater by night than by day
(c) be greater by day than by night.

(7) Which of the following bands of radio frequencies is known as MF:
(a) 30–300 MHz (b) 10–30 kHz (c) 300–3000 kHz.

(8) ACARS stands for:
(a) Atlantic crossing automatic reporting system
(b) Aircraft communication addressing and reporting system
(c) Air crew automatic reporting system.

(9) When the aircraft's No. 3 VHF is fitted with a VOICE/DATA switch, to use ACARS the switch position to select is:
(a) DATA (b) ON (c) VOICE.

(10) SATCOM uses INMARSAT satellites which are:
(a) polar orbiting (b) geostationary (c) equatorial orbiting.

Chapter 4
GROUND DIRECTION FINDING (VDF)

A ground station can be equipped for taking a bearing of an aircraft on that aircraft's transmission. In earlier days such a service operated in MF, HF and VHF bands. In the UK this service now only operates in the VHF band, that is, the normal communication band of 118 to 137 MHz. Service in other bands still exists in certain parts of the world and, where applicable, relevant information such as frequency in use, range and accuracy is found in the Aerad Flight Guide.

When the service is required, the procedure is to call up the station on the appropriate RTF channel. Frequencies are given in the COM section of the UK aeronautical information publication (Air Pilot) (UKAIP).

Services

A ground DF station can give true or magnetic bearings as follows:

QTE – aircraft's true bearing from the station
QUJ – aircraft's true track to the station
QDR – aircraft's magnetic bearing from the station
QDM – aircraft's magnetic heading to steer in zero wind to reach the station.

QTEs and QDRs are normally used in en route navigation as position lines. QDMs are requested when the pilot wishes to home to the station. QUJ is generally only known to academics.

In addition to the above, if you want a series of bearings or headings to steer (as for example you are *homing* to a station), the service you request is QDL. Where a three-station triangulation service exists (none in the UK at present, except on your MAYDAY or PAN calls), you may request a QTF. The station will take a fix on you and pass it to you in latitude and longitude co-ordinates or bearing and distance from a recognisable landmark, town, facility, etc.

Direction finding (DF) stations can refuse to give bearings if the conditions are poor or the bearings do not fall within the classified limits of the station. In this case, the controller will give his reason for the refusal.

Classification of bearings

According to the judgment of the operator the bearings are classified as follows. When the controller passes the bearing to the pilot he adds this classification to it, e.g. 'your true bearing 247°, class alpha'.

Class A – accurate to within ± 2°.
Class B – accurate to within ± 5°.
Class C – accurate to within ± 10°.
Class D – accuracy less than Class C.

Scope of the service

There are many automatic VDF stations whose purpose is purely to assist in radar identification for ATC purposes. These stations are not listed in the AIP for the obvious reason that they do not provide a normal DF service to the aircraft. The stations that are listed in the AIP provide normal 'homer' service and they are listed as such. Generally the class of bearing is not better than class B. Automatic VDF stations are not to be used as en-route navigation aids but their service is available to the fullest extent in case of emergency or where other essential navigation aids have failed.

Automatic stations (homers as well as those established for radar ident purposes) utilise a cathode ray tube for bearing measurements. With this type of equipment, the aerial is rotated in response to the incoming transmission and the direction of the received wave is displayed on the tube instantaneously. On the tube, the transmission appears as a trace and the bearing is read off against a scale. The main advantage here is that only a very short transmission is required – just long enough to read off the bearing.

Ranges available

Being VHF transmission, the range will primarily depend on the height of the transmitter and the receiver, that is, the line-of-sight range. Other factors, e.g. power of the ground and airborne transmitters, intervening high ground, etc., will also affect the range. Further, as explained in an earlier chapter, the aircraft may receive both the direct wave and the ground reflected wave in which case fading might be experienced or even the signals may be lost completely. But this situation would not linger too long and a satisfactory two-way communication would soon be restored.

Factors affecting accuracy

If an aircraft's transmission for a true bearing has been reflected by either uneven terrain or obstacles through its travel to the receiver or by objects on the site, the aerial will read a wrong direction. Thus, propagation error and site error will affect the accuracy. Poor accuracy is also obtained when nearly overhead the station, as there is a cone of no bearing.

The aircraft's attitude when transmitting signals may also affect the results. As you will remember, in general aviation, VHF communication transmission is vertically polarised. Best results are obtained when the transmission arriving at the ground DF aerial is vertically polarised. When an aircraft is in an attitude such that its transmission aerial is in the horizontal plane, the transmitted signals will be horizontally polarised and no signals will be received. In between the two extremes, poor reception on the ground may give poor results.

The effect due to coastal refraction on VHF is negligible.

VHF let-down service

The VHF let-down service, available throughout the world, has the primary advantage that the aircraft does not require any specialist equipment to carry out a let-down. The stations which provide this service are listed in the COM section of the UKAIP where you also extract frequency and the callsign. Details of the procedures are published by Aeradio and other aviation publishers. These details are also found in the RAC section of the AIP and it should be consulted whenever any details on terminal approach procedures (TAP) are not clear.

Two types of procedure are in current use: the VDF procedure and the QGH procedure. Generally, the VDF procedure is available and the UKAIP annotates in the remarks column against the station where QGH may be carried out. Where both procedures are available at the same station, the let-down pattern is usually the same.

For a VDF let-down, the pilot calls the station and requests VDF. The pilot is subsequently given a series of QDMs which he uses to achieve the approach pattern for landing, as published in the aerodrome landing chart.

With the QGH procedure the pilot is given headings to steer instead of QDMs. Based on the pilot's frequent transmission, the aircraft is first homed to the overhead (aerial) position at correct height. This height is the lowest available flight level or the safety altitude. When overhead the pilot will be given instructions for descent on the timed outbound leg. The aircraft will turn inbound on completion of this leg and further instructions to decision height will follow. Heading corrections will be given on the inbound leg until the pilot is in visual contact.

Test questions

(1) On a VHF let-down, the controller passes a true bearing of 127° class Bravo. The class Bravo means that the bearing is accurate to within:
(a) ±2° (b) ±5° (c) ±10°.

(2) VHF range for a VDF let-down is:
(a) $3 \times \sqrt{\text{Height (ft)}}$ (b) $1.5 \times \sqrt{\text{Height (ft)}}$ (c) line-of-sight.

(3) QTE means:
(a) aircraft's true track to the station
(b) aircraft's true bearing from the station
(c) aircraft's magnetic bearing from the station.

(4) When using VDF, errors:
  (a) are primarily due to coastal refraction
  (b) are nil because VHF transmissions are line-of-sight
  (c) occur mainly due to the homer siting.

(5) The Q-code for an aircraft's magnetic heading to steer to reach the station in zero wind is:
  (a) QDM (b) QDR (c) QTE.

Chapter 5
AUTOMATIC DIRECTION FINDING (ADF) AND NONDIRECTIONAL BEACONS (NDBs)

In navigation, when plotting radio bearings, care is required in deciding who does the work; the pilot in the aircraft or the operator on the ground. Here we deal with the aircraft getting a bearing on a ground-based radio station. Ground radio stations providing such a facility are known as nondirectional beacons (NDBs).

NDBs transmit vertically polarised signals in the MF/LF band. They radiate equally in all directions, hence their name. An aircraft carrying associated ADF equipment can tune in to a station on its published frequency and can measure the direction of the incoming ground waves. The datum for the direction measurement is taken from the nose of the aircraft and therefore, the ADF indications are *relative bearings*. On modern equipment these bearings are displayed automatically (ADF) and when fed to a radio magnetic indicator (RMI), QDMs are indicated. Additionally, on some automatic equipment, a facility is provided to obtain bearings manually and to check the sense.

A loop aerial is used in the aircraft to determine the direction of the ground transmitter. In an aerodynamically smooth aerial unit, it is made up of a number of strands of wire (to give them an ideal aerial length) wound round a frame, and is mounted in the most suitable position on the fuselage. The aerial itself may be rotatable or fixed, depending on the type of equipment, but modern installations usually have integrated ADF loop/sense antennae with a single cable coupled to the flight deck. Such models have a fixed cross loop system with a motor-driven goniometer in the receiver. These have the advantages of allowing an almost flush aerial mounting and all of the moving parts within the receiver.

Loop theory – rotatable loop

The vertical members of the loop are designed to pick up the signals. When the plane of the loop is parallel to the direction of the vertically polarised NDB radiation, signals will be picked up by the two vertical arms. There will be a phase difference between the signals arriving in the two arms because of the distance between them. This will cause a current to flow in both arms. These signals in the two arms are in opposition round the loop and therefore,

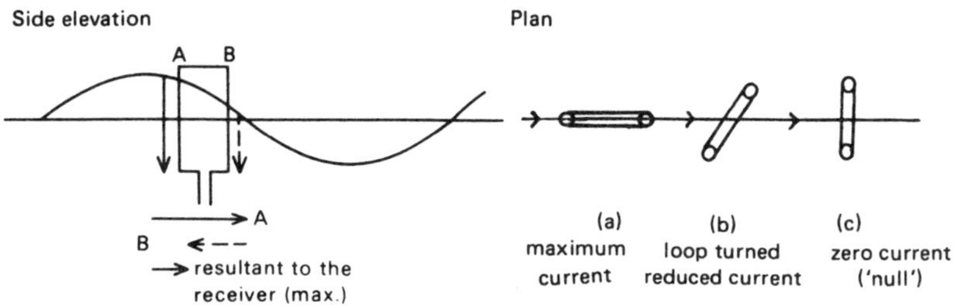

Fig. 5.1 Loop aerial.

the signals (or voltage) finally passed to the receiver will be the difference between the two, see Fig. 5.1. Notice that with the loop in position (a) the distance between the two arms is at a maximum. Therefore the phase difference and the resultant current flow to the receiver is also at a maximum.

If the aerial is now rotated through 90°, both arms will face the transmitter together and the incoming wave will reach them at the same instant, that is, at the same phase (Fig. 5.1 position (c)) and the resultant current flow is nil.

Thus, as the loop is rotated from position (a) to position (c) the current flow is reduced from maximum to nil. Therefore at any intermediate angle (b) the current flow is a function of the cosine of the angle the aerial makes with the incoming wave.

If the loop is further rotated beyond 90°, current will start flowing again, but in the reverse direction. If a polar diagram is traced out showing the signal strength produced by the loop at different angles through 360°, the result is a *figure of eight*, see Fig. 5.2.

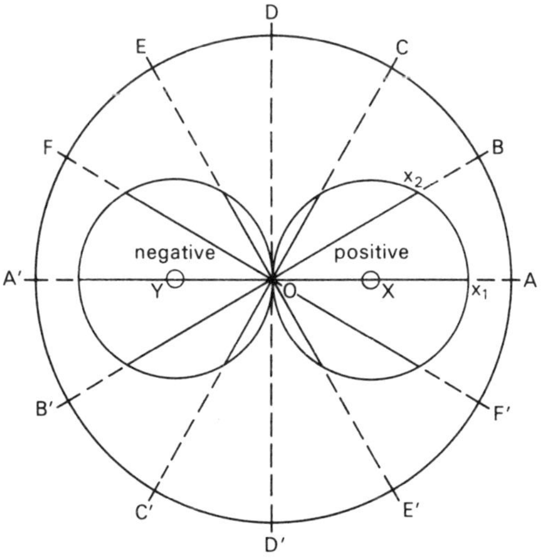

Fig. 5.2 Polar diagram.

In Fig. 5.2, XY is the loop, with its centre at O, and A to F and A′ to F′ are various positions of an NDB at 30° intervals. With the transmitter in position A, the loop is lying parallel to the incoming wave, X end leading, the current produced is at a maximum. Plot this as a vector, one unit in length, distance Ox_1 in Fig. 5.2.

In position B, the current induced is shown by vector Ox_2 and its value is 0.86 vector units. In position D, cosine of 90° is 0, and when the transmitter is at E, cosine of 120° is −0.5. As you go through 360° you will have traced out a figure of eight, half of it will be positive and the other half, negative.

The polar diagram has two positions where maximum signals are being received (A and A′) and two positions where no signals are being received (D and D′). The zero strength positions are known as null positions.

When establishing the direction of the ground station, the loop can be aligned so that the plane of the loop is parallel to the incoming signal, that is, the loop-transmitter relationship as in position A. This is done by turning the loop until maximum signal strength is heard, and calling that the direction of the station. The disadvantage of ascertaining the station direction from maximum signal is that the field strength on either side of the maximum falls very slowly and consequently the determined direction can be in error.

For this reason the null position is used to determine the transmitter direction. Nearer 90°, (loop-transmitter relationship as at D) the value of cosine falls rapidly and a total null is more easy to recognise than a total maximum. However, it should be noted that just as a total maximum may occur at A or A′, the null may occur when the transmitter is in the D or D′ position. In other words, there is a 180° ambiguity in the bearing indication.

This is resolved by the use of a sense aerial. It is an omni-directional aerial and its polar diagram is a circle, the radius of which is electronically adjusted to fit on top of the figure-of-eight polar diagram. To resolve the ambiguity these two fields are mixed together. The resultant polar diagram is a heart-shaped figure, called a cardioid (Fig. 5.3). It will be noticed that a cardioid has only one null position, even though it is somewhat 'blunter' than the null positions appropriate to D and D′ of Fig. 5.2.

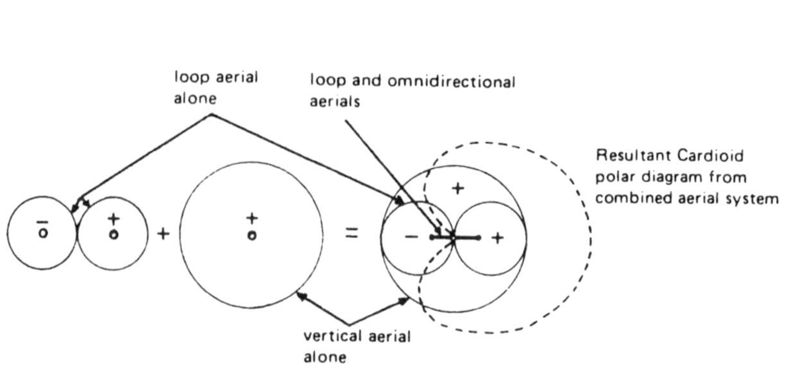

Fig. 5.3 Cardioid polar diagram.

Loop theory – fixed loop

In the Bellini–Tosi method of direction finding, the system uses an omni-directional sense aerial and a pair of fixed loops at right angles to each other: one loop with its axis in the fore and aft plane, the other in the athwartships. These loops are connected to the field coils (stator) of a goniometer and the current in these coils sets up a field about a rotor. The direction of the field is related to the direction of the incoming signal.

If the rotor coil lies at 90° (null position) to the resultant magnetic field in the stator, the system is in balance. When the rotor is not at the null position, a current is induced in the rotor. This activates a motor which drives the rotor the shortest way to the null position.

Resolution of ambiguity

When the equipment is functioning on ADF, 180° ambiguity is automatically resolved for you. In one system a pattern of 'switched' cardioids is produced by combining the output of an omni-directional sense aerial and alternating the connections rapidly.

The signal output from the cardioids governs the voltage applied to two uni-directional motors geared to the loop. The left-hand cardioid influences one motor, the right-hand cardioid the other. The motors turn in opposite directions and by means of the differential gear they rotate the loop in the appropriate direction.

When the loop is in the equi-signal position, the differential output is zero and the pointer is steadily indicating correct direction. This steady indication has a slight hunt about its mean position. This has two benefits. It ensures that the ambiguity cannot exist; if the pointer was indicating 180° out, as the needle departed from its mean position a voltage would be induced in one of the motors and would drive the needle through 180° to the correct null. The other benefit is that the steady movement of the needle about a mean position is an indication that the ADF is working and a good bearing is being displayed. Excessive hunt would develop when the incoming field strength is

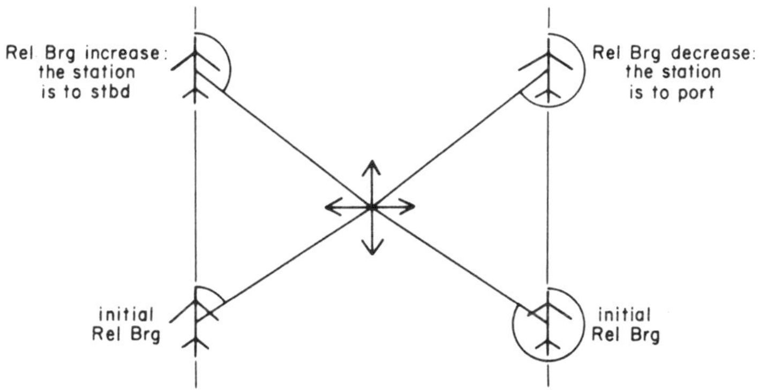

Fig. 5.4 Resolution of ambiguity.

weak or the signal/noise ratio is low. With experience the operator is able to assess the degree of accuracy of the indications.

Occasionally, in conditions of heavy static or interference from a powerful transmitter on a nearby frequency it might be advantageous to operate on manual loop if this facility is available on the equipment. The loop aerial is turned manually until it arrives at the null position, but the indication must then be checked for sense.

Ambiguity may be resolved from knowledge of the DR position. You can also resolve it by observing changes in bearing with time. Unless you are homing directly to a station (or away from it) the bearings increase or decrease as the flight progresses. Study Fig. 5.4 which is self-explanatory.

Coverage of NDB transmissions

The rated coverage of NBDs in the UK is noted in the COM section of the UKAIP. Within this area the field strength of the wanted signals to unwanted signals exceeds the minimum value specified. ICAO recommends this value to be 15 dB but no less than 10 dB, a 3:1 protection ratio. This protection takes into consideration the average atmospheric noise in the geographical area concerned but not the presence of sky waves at night.

The rated coverage of an NDB depends on the frequency, transmission power and the conductivity of the path to the coverage boundary. These three factors, being measurable, enable the planners to establish the facilities by controlling the output power and allocating frequencies. The lowest field strength (ICAO recommendation is 70 microvolts/metre in Europe) will be received at the boundary.

Types of NDBs

Locators. These are low powered NDBs in the LF/MF band, usually installed as a supplement to ILS and located at the sites of the outer and middle markers. A locator has an average radius of rated coverage of between 10 and 25 nm. The type of emission is N0N A2A and they identify by a two or three-letter morse group at seven words per minute once every ten seconds. During the ident period the carrier remains uninterrupted to ensure that the pointer does not wander away. The ident is effected by on/off keying of the amplitude-modulating tone.

Homing and holding NDBs. These are intended primarily as approach and holding aids in the vicinity of aerodromes, with rated coverage of around 50 nm. The class of emission and method of ident are similar to the locators.

En-route and long-range NDBs. These provide en-route coverage along the airways and a long-range bearing facility for ocean tracking and similar operations. The NDB at Cocos Island may be heard from several hundred miles.

The recommended type of emission is N0N A2A unless the required rated coverage is not practicable because of interference from other radio stations or high atmospheric noise or other local conditions, in which case N0N A1A emission may be used. With this type of emission the identification is effected by on/off keying of the carrier (during which period the pointer may wander away), and the NDB identifies at least once every minute.

The amplitude-modulating signal causing identification for N0N A2A emission is either a 1020 Hz or a 400 Hz tone.

The choice of frequency band

The requirement is to produce surface ranges of intermediate order.

Frequency bands from VHF and above, being line-of-sight propagation, may be ruled out from consideration. HF would produce a very short ground wave, and the sky waves would interfere with the ADF operation day and night. VLF installation and running costs are high and require large aerials. Static noise is excessive.

The frequency bands chosen are upper LF and lower MF. In these ranges, aerial requirements are acceptable, static is less severe than VLF, there is no interference due to sky waves during the day time and the bands are ideally placed to produce the required ranges.

Frequencies

Although the frequencies are allotted from LF/MF, by convention an NDB is an MF aid. Frequencies assigned by ICAO are from 200 kHz to 1750 kHz. In the UK and Europe, NDB frequencies are normally found between 255 kHz and 455 kHz.

Functional checks

ICAO provides for monitoring of NDBs for radiated carrier power (not allowed to fall below 50%), failure of identification signals, failure of the monitor itself or other malfunctioning. In the UK the NDBs are not regularly monitored but functional checks of NDB and locators are carried out at eight hour intervals during the period of service.

Automatic direction finder (ADF)

The ADF is a radio receiver that is able to identify the direction from which a signal is being received. It uses a loop aerial to determine the direction and a sense aerial to resolve the 180° ambiguity. An aircraft equipped with ADF may 'home' to the transmitter or use it as an aid to navigation. The relative bearings are displayed on the relative bearing indicator (RBI) and QDMs are indicated on an RMI.

Although there are many varieties of ADF, which used to be called 'radio compass', on the market they are all basically the same. The main components of the system are:

- a radio receiver operating in the LF and MF bands
- a control box
- a shielded fixed or rotatable loop aerial
- a non-directional aerial
- one or more indicators.

Typical control units are shown in Figs. 5.5 and 5.6.

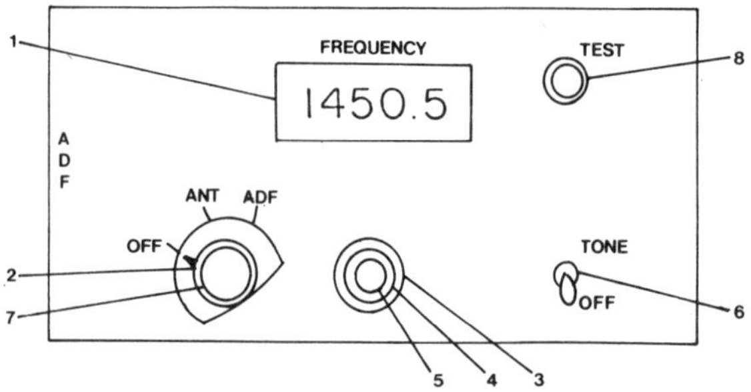

1: Illuminated dial indicating frequency selected. 2: Mode control switch (and power-off control). 3: Frequency control knob (in 100 kHz). 4: Frequency control knob (in 10 kHz). 5: Frequency control knob (in 1 and 0.5 kHz). 6: Tone ON-OFF switch. 7: Audio GAIN control. 8: Built-in self test control.

Fig. 5.5 ADF control unit.

Controls

The ADF control units illustrated in Figs. 5.5 and 5.6 cover just about every type of control on ADF equipment. Not all ADFs have all the same facilities.

Function switch. The number of positions for a function switch varies with the equipment. The purposes of the various positions are as follows:

(1) *Off position.* This is the normal position when the ADF is not in use.
(2) *ANT.* In this position the sense aerial is in the circuit and tuning of the station is carried out in association with the frequency band selector and the selectivity switch(es). This switch is variously named as REC (receiver), ANT (antenna), OMNI or SENSE.
(3) *ADF.* In this position both sense and loop aerials are in operation and the ADF gives a continuous indication of the bearing of the station to which the receiver is tuned. This position is also known by other names, e.g. COMP.

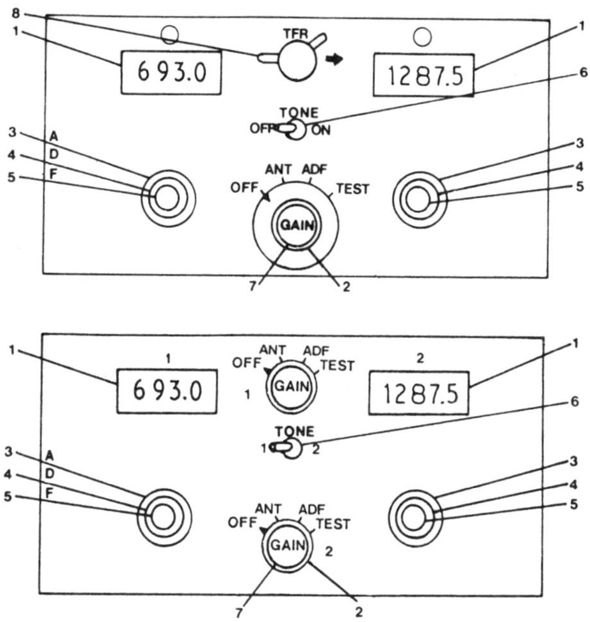

1: Frequency indicating dial. 2: Mode control switch. 3, 4, 5: Frequency selector (100, 10, 1 and 0.5 kHz). 6: Tone selector. 7: Audio GAIN control. 8: Transfer (TFR) switch – selects one of the two chosen frequencies and illuminates appropriate indicator.

Fig. 5.6 ADF control unit.

TONE or BFO or CW/RT or RTF. The need for this facility arises because the NDB emission may be N0N A1A or N0N A2A. It will be recalled that in N0N and A1A, parts of the emission are at a radio frequency which cannot be heard, being beyond human ear range. When the CW/RTF switch is in the CW position or the TONE or BFO switch is in the 'ON' position, these emissions are made audible. This is done by the use of an oscillator inside the receiver which produces internally a frequency slightly different from the frequency being received from the NDB. For example, if the received frequency is 400 kHz, it may produce 399 kHz internally. It then takes the difference between the two (1 kHz in our example) which is an audio frequency. This difference is called the beat note and hence the name BFO – beat frequency oscillator.

Commercial broadcasters may sometimes be used to take a bearing. In this case, since the transmission is in audio frequency the BFO is not required.

In N0N A1A emission, N0N is the carrier part and it is transmitted for about 10 seconds; it is then interrupted to give the A1A part, which is the ident and may last for around 5 seconds. During the ident breaks the pointer may wander away. For this reason N0N A1A emission is not in favour with ICAO and the long range NDBs generally give idents at longer intervals to minimise breaks.

Similarly N0N A2A emission is again an unmodulated carrier for about ten seconds but the ident part, A2A, is an amplitude-modulated carrier. The ident lasts a few seconds and generally the carrier is not broken during ident period.

Filter. If fitted, this position is used to reduce the amount of spurious background noise.

Uses of ADF

(1) As an en-route navigation aid, position lines may be obtained. By taking two or three bearings on the same or different NDBs, fixes are obtained.
(2) Flying airways – there are many airways in the world still marked by NDBs.
(3) A fix is obtained when overhead an NDB – useful on airways for position reporting.
(4) An NDB can be used for holding at an en route point or at the destination aerodrome, for homing to the station and carrying out a let-down.

Procedures for obtaining bearings

(1) Check that the aircraft is within the promulgated range of the station to be used
(2) Check the frequency and callsign
(3) If available, use the pre-select facility
(4) Tune the station
(5) Identify the station
(6) Function switch to ADF
(7) Note the bearing being indicated, and take the time; this is the time of your position line, and the bearing obtained is a relative bearing
(8) If the bearing is required on an RMI, select NDB on red (thin) needle and the bearing indicated is a QDM
(9) Unless planning to take a further bearing in a short space of time, return function switch to ANT or even switch the ADF off.

Manual determination of bearings

On occasions when excessive interference is present, the pilot may find it advantageous to obtain a bearing aurally by manual operation of the loop aerial. Not many modern aircraft ADFs will have this facility but where the facility exists, the procedure is as follows:

(1) Steps (1) to (5) as above
(2) Select LOOP position on function switch
(3) By operation of the loop L/R control, rotate the loop until the signal strength heard in the earphones is at a minimum (you are trying to locate the 'null' position)
(4) The bearing is being indicated at this stage, but it is necessary to check the sense unless it can be resolved by one of the other methods; press SENSE button and rotate the loop clockwise by holding the loop control to the right
(5) If the signal strength in the earphones decreases, the bearing is correct (cardioid's null is approaching you)
(6) If the signal strength increases, swing the loop through 180° and repeat the sequence.

An illustration of relative bearings and indications on an RBI is shown in Fig. 5.7.

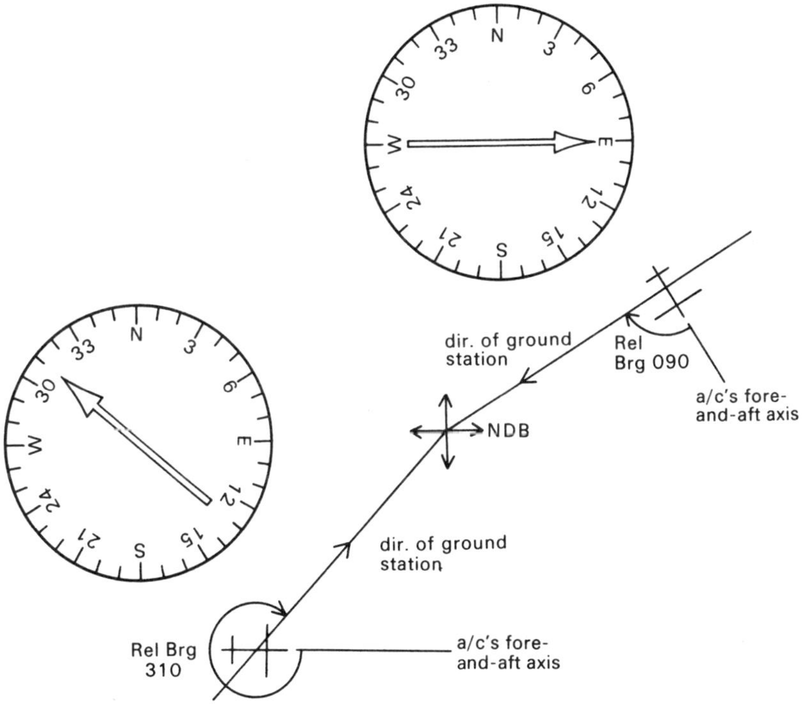

Fig. 5.7 RBI indications.

Homing to a station

When homing to a station on the ADF, the point to bear in mind is that the station is directly ahead of you (if not, you want it to be so) and therefore, the ADF indication is 360° or around 360°. Similarly, if you are leaving a station on the ADF, the station is directly behind you and the indication you are looking for is 180° or around 180°.

Theoretically, a station may be reached by maintaining 360° on the ADF, but if there is any wind blowing (and when doesn't it blow?) you will fly a curved path to the station. Further, consequent upon wind, you will be continually altering heading to combat displacement caused by the wind and ultimately you will arrive overhead from a direction facing into the wind. See Fig. 5.8.

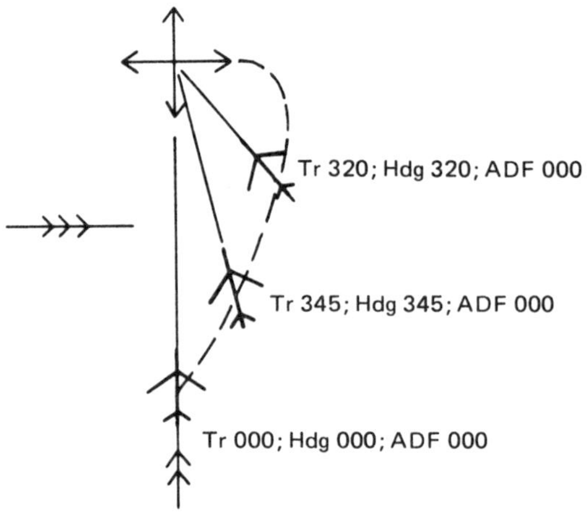

Fig. 5.8 Maintaining ADF 000°.

The above method of homing, apart from being time-wasting, may not be possible due to track maintenance requirements. In congested areas and on airways, when homing from beacon to beacon you are required to fly notified tracks. This may be done by simply making allowance for the wind velocity right at the start, adjusting as the flight progresses if a wind change is noticed.

Suppose the drift is 10°S at the start (Fig. 5.9). Steer 350° to allow for the drift. The ADF will then read 010° (take away from the nose, add to the indication). As long as 010° remains indicating, you are maintaining the track, and you will arrive overhead. If the indication starts a gradual decrease, you are drifting to starboard, and you must allow for more than 10°S drift.

For the same reasons, if the reading starts to increase, you have allowed for too much drift and should alter the heading accordingly. After one or two such alterations you will hit on a heading which is right.

If while juggling with headings you managed to get off the track, or if you wish to join a given track, the technique employed is to intercept the desired track at a convenient angle, generally 30° – that is, at an angle of 30° between interception heading and the desired track. To calculate what bearing the ADF will indicate when you arrive at the track, the rule is:

Add on the nose (i.e. heading), take away from 360;
Take away from the nose, add on 360.

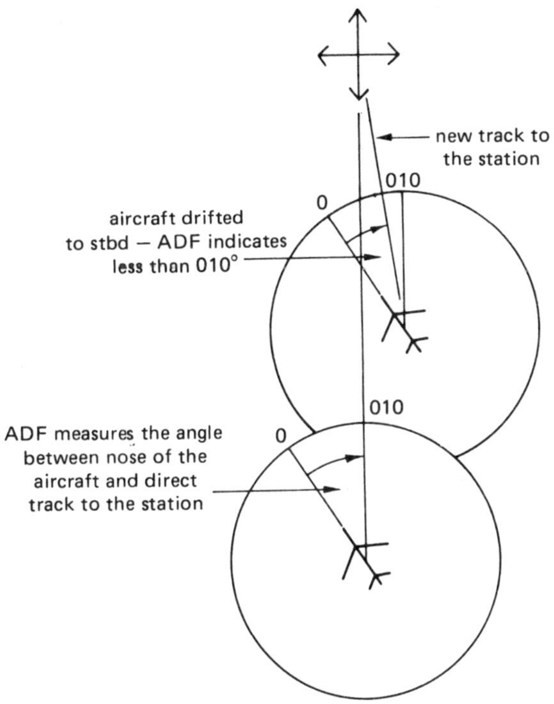

Fig. 5.9 Tracking directly to beacon.

Thus, if you intercept at an angle of 30° and your heading is smaller than your track (that is, you have taken away 30° from the nose) ADF indication of 030 will tell you that you are crossing the track.

An aircraft in position 1 in Fig. 5.10 wishes to intercept and follow a track of 070°; drift 5°P. Follow stages 1, 2 and 3.

Tracking away from the station

As the station approaches, a rapid buildup in volume is noticed; when passing the station, signals will momentarily fade, followed by another rapid increase in volume indicating station passage. The station passage is also indicated by increasing oscillations of the needle, subsequently settling down to an indication near 180°. When tracking away, the relative bearing indicated will be greater than 180° if starboard drift is being experienced (Fig. 5.11), and less than 180° if port drift is being experienced. The procedure is to fly out with the drift applied to maintain the track. Then you know what ideal bearing you are looking for. As the flight progresses if the aircraft is not tracking correctly, the pointer will start drifting slowly to one direction or the other. If the readings are increasing, you are experiencing starboard drift; port drift if they are decreasing – the same rule as stated earlier.

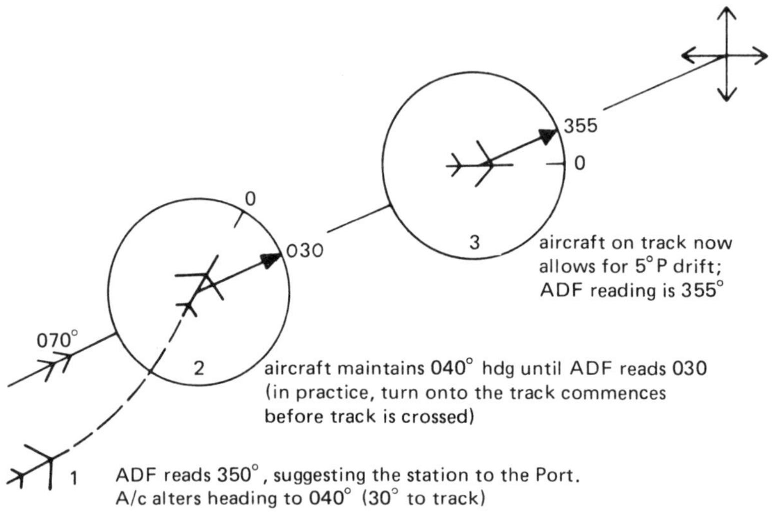

Fig. 5.10 Homing on ADF.

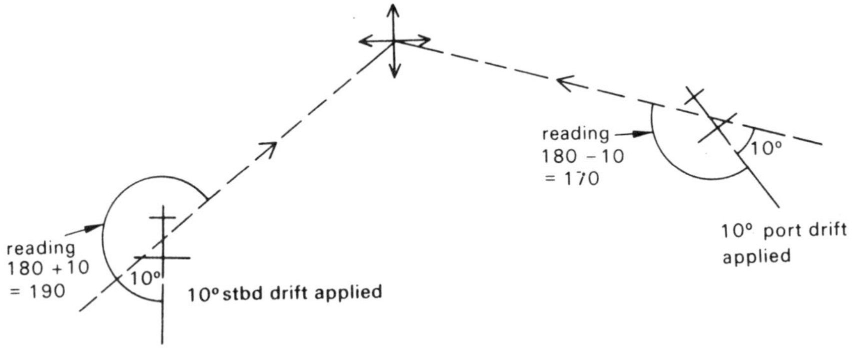

Fig. 5.11 Tracking away from station.

Holding patterns

Generally the holding patterns are race-track type patterns and all turns in the pattern may be either right-hand (RH pattern) or left-hand (LH pattern). This is indicated on standard terminal arrival route charts (STARs) and they are studied carefully as a matter of pre-flight preparation. Holding procedures are carried out in two phases: entering the pattern and subsequently, holding in the pattern. To enter the pattern, depending on the direction of approach to the NDB, it may be necessary to carry out a pre-entry manoeuvre. 360° approach directions round a holding NDB (as well as any other holding facility) are divided into three approach sectors as shown in Fig. 5.12. Each sector has its own procedure to get you in the pattern.

To clarify the sector arrangements, let us examine the RH pattern in Fig.

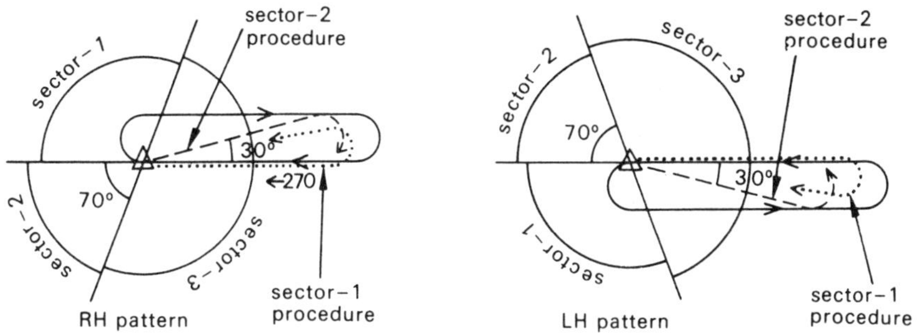

Fig. 5.12 Holding patterns.

5.12. The inbound track is 270° (M). The sector divisions based on this track are as follows:

Sector 1 when approaching NDB between 090° (M) and 200° (M).
Sector 2 when approaching NDB between 090° (M) and 020° (M).
Sector 3 when approaching NDB between 020° (M) and 200° (M).

If your approach track is one of the dividing lines you may choose either sector. Procedures to be adopted in individual sectors in an RH pattern, are as follows:

Sector 1

(1) On arrival overhead, fly parallel to the reciprocal of the inbound leg for the appropriate time.
(2) Then turn left and home back to the NDB.
(3) On second arrival over the facility turn right and commence the pattern.

Sector 2

(1) On arrival overhead, make good a track 30° to the reciprocal of the inbound leg towards the inside of the pattern.
(2) At the appropriate time, turn right and join the pattern on the inbound leg, and home to the facility.

Sector 3

On arrival overhead, join the pattern directly.

Actual holding at the facility commences now. Turn right through 180° rate one and start timing when abeam the NDB. How do we know when we are abeam? Well, in zero wind conditions, an RBI pointer indicating 090°R indicates abeam position. To this figure add the amount of drift if it is starboard, subtract from it if it is port. In Fig. 5.13 the aircraft having a 10°S drift would wait for its ADF to indicate 100°R.

The outbound track is parallel to the inbound track. Apply drift and adjust

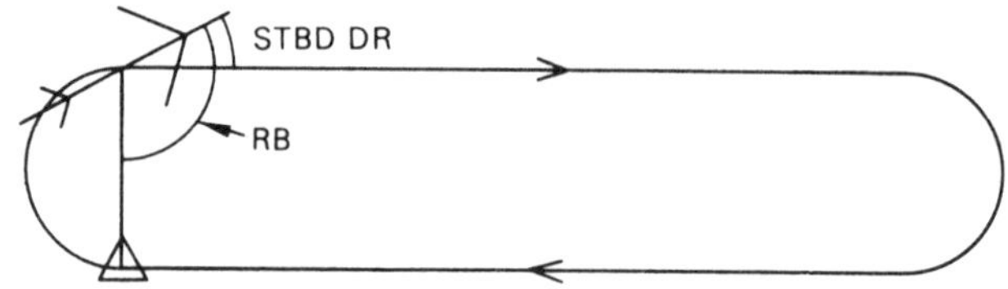

Fig. 5.13 Race track.

the leg timing for your ground speed to maintain the track. At end of the leg, turn right through 180°, rate one and fly inbound track by ADF.

NDB let-down

On a terminal approach procedures chart, a typical let-down looks something like Fig. 5.14. The let-down pattern is given in both plan and elevation and at

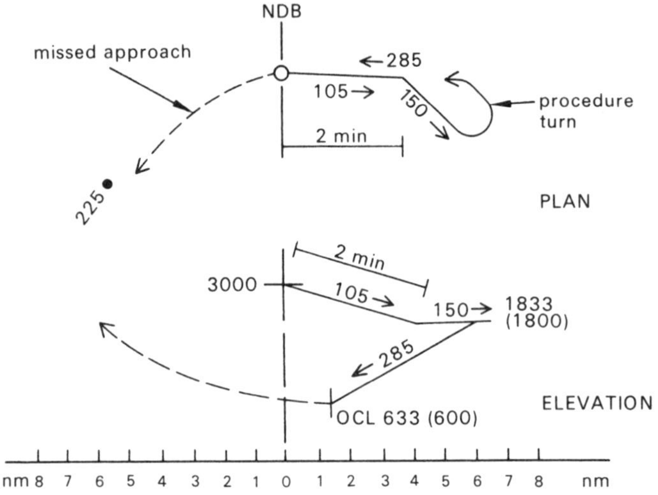

Fig. 5.14 NDB let-down.

the bottom of the chart there is usually a scale showing distances in nautical miles from the NDB position. All bearings shown on STARs are magnetic. A procedure turn may be a level turn or a descending turn. The outbound track may not be the reciprocal of the inbound track. In this case, a rate one turn through 180° generally brings the aircraft on to the inbound track. This turn, again, may be a level turn or a descending turn. Rate of descent on all approaches, unless otherwise shown on the STAR, is 650±150 ft/min. At certain aerodromes 'shuttle' procedures are available permitting an aircraft to descend to the altitude from where a let-down may commence. Heights given on a STAR are related to aerodrome elevation with QFE equivalents in brackets.

During a let-down, maintain tracks by applying drift and flying with reference to ADF. Where a leg is time-controlled, adjust that time to compensate for your ground speed. A procedure turn is a 45 second leg followed by a turn on to the inbound track.

A careful study of all possible destination and alternate let-down procedures should be made as part of the pre-flight preparation. As many details as possible are extracted in advance and arranged in a convenient form so that you are ready to commence the procedures at your destination with a minimum of work at that end. In case of any doubt in interpretation of information in TAP, the UKAIP should be consulted. The following information is needed for an NDB let-down:

Height	Track (TR)	Drift	Ideal Brg on ADF	True airspeed (TAS)	GS	Dist.	Time	Rate of descent (ROD)

Finally, a pilot must not descend below his decision height (DH) unless the conditions are equal to or better than those specified in the operations manual.

Angle of lead

A turn on to the track must be commenced before the track is reached, otherwise the track will be overshot and a further alteration of heading in the opposite direction will be necessary. This angular allowance that you will make for the turn is known as the 'angle of lead'. It is dependent upon various factors, e.g. aircraft's TAS which governs its radius of turn, its distance out from the station, wind velocity and the aircraft's rate of turn. Two points must be noted:

(1) For a given airspeed, the angle of lead decreases as the distance from the station increases (Fig. 5.15).

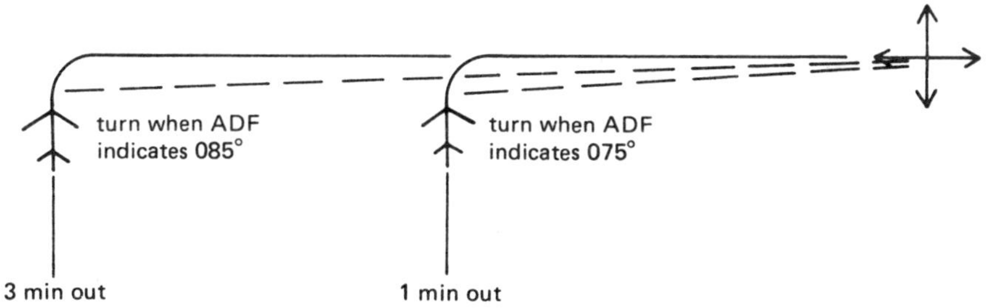

Fig. 5.15 Angle of lead – fixed TAS.

(2) For a given time out from a station, the angle of lead is constant irrespective of aircraft speed (Fig. 5.16).

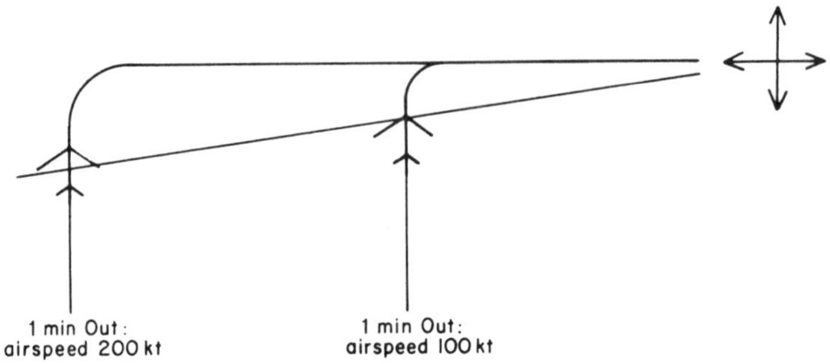

Fig. 5.16 Angle of lead – fixed time out.

Factors affecting range of NDBs

The undermentioned factors affect the useful ranges available from an NDB.

(1) ***Transmission power.*** The range at which an NDB may be successfully used depends on its power output. Being an MF transmission, an increase in power means an increase in the range available. The power requirements are, however, such that to double the range, the power must be increased fourfold (or, the range is proportional to the square root of the power output). It must also be remembered that the power output of a station is strictly limited to the value that will produce acceptable field strength at the rated coverage boundary. Associated with the power output, the radiation efficiency of the transmitter aerial (e.g. height and other characteristics of the radiating system) governs the field strength of the signals received from an NDB.

(2) ***Frequency.*** The operating radio frequency is a factor which also governs the field strength of a signal being received. As we learnt earlier, for a given transmission power, the lower the frequency, the lesser the attenuation and the greater the ground wave range.

(3) ***Type of terrain.*** This factor reduces the useful ranges in two ways: by affecting the field strength of the signal and by giving inaccurate information.

   (a) *Type of surface.* The conductivity of the path between the NDB and the receiver determines the attenuation of the signals, and the field strength. Longer ranges are always obtained over water than over dry soil.

   (b) *Mountains and other obstructions.* Mountains and other obstructions may block the signals, or more likely the signals will be received after having been reflected from peaks and valleys or due to diffraction and scattering effect. These signals, not necessarily arriving from the direction of the station, will give erroneous indications. The effect is more pronounced at low levels; a gain in altitude is required to minimise the effect.

(c) *Coastal refraction.* In coastal areas the differing radio energy absorption properties of land and water cause refraction of radiated waves. The error in the indications is caused by what is called coastal refraction. If a wave does not leave the coast at 90°, it will bend towards the medium of high density, that is, landmass (Fig. 5.17). The amount of refraction depends on the angle between the signal and the coastline. Errors of the order of 20° may occur when the bearing is taken on a signal leaving the coast over 30°. Thus, this factor reduces the effective ranges through giving erroneous readings. To minimise the error, use an NDB sited on the coastline or climb up to a higher flight level or use the signals which leave the coast around 90°. (However none of these is a very practical solution.)

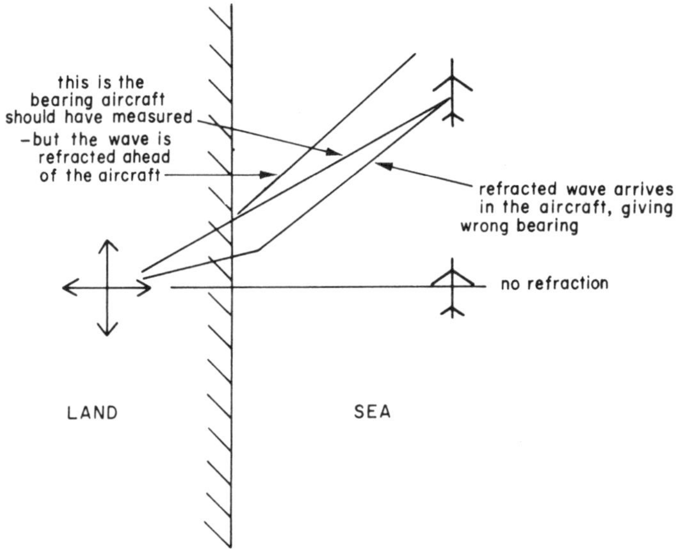

Fig. 5.17 Coastal refraction.

(4) ***Night effect (sky wave error).*** The effective range of a long range beacon which has a daytime range of 200 nm will be reduced to about 70 nm by night. A serious cut-down in the range occurs due to the presence of sky waves in the LF/MF band. In this band during the day time, sky waves are not normally present, but at night they affect the ADF accuracy when they enter the horizontal members of the loop from above (Fig. 5.18).

These waves will start coming from skyward from approximately 70 to 100 nm range upward. If the receiver is also within the range of the ground wave, the two signals will mix and distortion of the null will occur. On an ADF, this will be indicated by excessive oscillations of the needle. True null

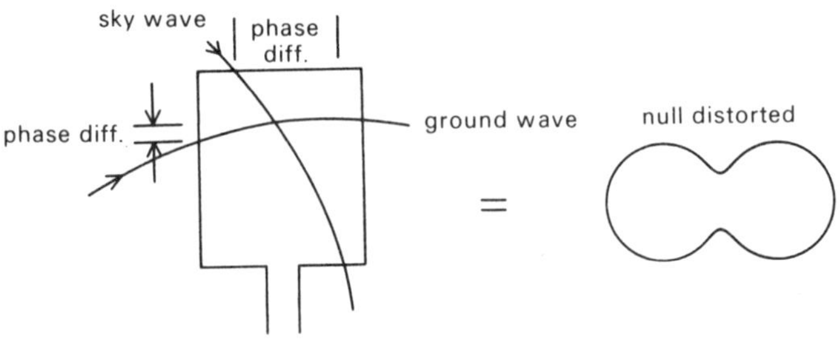

Fig. 5.18 Night effect.

is masked and efforts to find a mean over a wavering area can lead to error since the mean is not necessarily the centre of a wavering needle. If the aircraft is only receiving sky waves, there will be a good null but that indication can again be erroneous since there is no guarantee that the reflecting plane is parallel to the earth's surface. In other words, the signal may not arrive from the station direction.

This effect on the ADF is called night effect. It is minimised when using a station in the lower section of the frequency band, thus reducing the incidence of the sky wave. Dusk and dawn are critical periods for ADF operation – extreme care must be exercised. Also, use a more powerful beacon if you have a choice. Lastly, choose the station nearest to you.

The presence of sky waves is indicated to the operator in two ways: excessive oscillation of the needle, and fading of the signals.

(5) ***Protection range.*** Irrespective of the range that an NDB is capable of producing, the use of an NDB in the UK is restricted to ranges promulgated in the UKAIP. This is known as protection range and this restriction is necessary in order to provide reception, free from interference from other NDBs transmitting on the same or similar frequencies.

The main reason for the interference is congestion in the MF band. In Europe (and in America) there is always a heavy demand for channel space in this band, which is already overcrowded with NDBs and broadcasters. In LF and MF (lower end of the frequency spectrum) interference between two *similar* frequencies can take place quite easily (a common experience while tuning radio stations at night) and the protection is afforded by the Civil Aviation Authority (CAA) controlling the allocation of power and frequency to the stations. Protection ranges are based on providing a minimum protection ratio of 3. That is, the ratio of the field strength of the wanted signal to unwanted signal will be at least 3 to 1. Or, when translated in terms of noise, the level of wanted signal will be at least 10 dB higher than that of unwanted signal within the promulgated range. In the UK the protection ensures that the errors in the service area will not exceed ±5°. But this protection is guaranteed only during day time. At night:

(a) Sky waves seriously affect the operation of the ADF due to night effect.

Further, the E layer gains height at night, increasing the ranges where sky waves could be received. In other words, the sky wave from a distant NDB operating on a similar frequency will extend its range and produce interference within the protection range of your NDB.

(b) Broadcasters and other high-powered radio beacons will gain field strength at night. This means that at a given place, where during the day time you would receive wanted signals at a level of at least 10 dB higher than the unwanted signals, the unwanted signals now increased in field strength will produce higher noise, and the protection minimum will fall.

(c) A forecast of interference-free ranges could not be made for night periods since the height of the ionospheric layer is variable and its density is also variable. Among others, these two factors decide the range of the return of the sky waves.

Therefore, the principle of protection ranges breaks down at night and the useful ranges are greatly reduced. Extreme caution should be exercised when using the ADF at night and it is most important to ensure that correct tuning is done to the exclusion of any unwanted signals.

(6) ***Static.*** All kinds of precipitation (including falling snow) and thunderstorms, together with solid particles such as dust in the atmosphere can cause static interference of varying intensity to ADF systems. Precipitation static reduces the effective range and accuracy of bearing information. Thunderstorms can give rise to bearing errors of considerable magnitude, even to the extent of indicating false station passage.

It is not at all unusual for the pointer to point in the direction of a thundering cumulonimbus cloud.

The extent to which a receiver will admit the noise depends on:

(a) the bandwidth of the receiver
(b) the level of the atmospheric noise
(c) the level of noise produced by other interference (e.g. other radio stations, industrial noise).

(7) ***Type of emission.*** For a given transmission power, the ranges produced by an A2A emission are shorter than a A1A emission.

Factors affecting accuracy of ADF

Some of the factors mentioned above also affect the accuracy of the indications. There are other factors which affect accuracy without affecting range directly. Factors affecting accuracy are:

(1) ***Night effect.*** (Already discussed.)
(2) ***Type of terrain.*** Mountains, physical obstructions, refraction on leaving the coast.
(3) ***Static interference.*** The cause of static has already been discussed.

Static can affect the accuracy of ADF at all ranges. At a comparatively short range, e.g. less than 50 nm, static and other noise is considered to be potentially dangerous and in these conditions the indications should be monitored on a VHF aid if available.

(4) ***Station interference.*** This is another potentially dangerous situation at short ranges. When two stations at different locations are transmitting on the same or similar frequency, the bearing needle will take up the position which is the resultant of the field strengths of the two transmissions. The indications may give large errors. To avoid this, use the facility when you are inside its promulgated service area.

(5) ***Quadrantal error (QE).*** This error occurs because the fuselage reflects and re-radiates the signals hitting it. These signals mix with the signals entering the loop, giving an error in the indication. Signals arriving at the loop from the aircraft's relative cardinal points are not affected. Signals hitting the fuselage at any other angles are affected, the maximum effect is noticed when the signals arrive from the direction of the aircraft's relative quadrantal points. Hence the name quadrantal error. These signals are bent towards the aircraft's major electrical axis, which is normally its fore-and-aft axis.

An ADF is regularly calibrated and corrected and any remaining errors are recorded on a QE card. With modern equipment and improved techniques this error is of academic interest only.

(6) ***Loop misalignment.*** If the loop aerial is not exactly aligned with the fore-and-aft axis of the aircraft all bearings subsequently measured by the equipment will be in error by the amount of misalignment. This error is eliminated by careful fitting and aligning of the loop.

(7) ***Lack of failure warning device.*** Because there may be no cockpit indication of a ground or airborne equipment failure, a serious situation can arise if the pilot continues to follow a steady indication when in fact no information is being fed to the pointer. This can occur for a variety of reasons: the pilot throwing the function switch to standby position through habit or absentmindedness, the ground transmitter stopping radiation, airborne receiver going unserviceable and so forth. The only way to prevent such an incident, in the absence of a warning flag or other built-in indication, is to constantly monitor the identification signals.

Effect of aircraft height on range

NDBs transmit in the LF and MF bands. In these bands the radio waves curve with the surface of the earth. Because of this, an aircraft will receive signals, if otherwise within the range, no matter how low the aircraft is flying.

Since these waves also propagate in space, aircraft at higher altitudes will also receive the signals. Thus, the height of an aircraft has no significance as far as the range is concerned. Height, however, may become significant as pointed out earlier, when flying in mountainous areas or using coastal NDBs.

Signals/noise ratio

To improve signals/noise ratio:

(1) narrow down receiver bandwidth ensuring that you do not cut out your own signals.
(2) tune the wanted station carefully, and
(3) if it is thought that the unwanted signal is the result of a more powerful beacon encroaching on your bandwidth, see if you can exclude it by off-tuning your own station.

Some interesting facts about ADF

(1) If there are two NDBs, one on the coast and the other fairly inland, and if the coastal refraction for both propagations is the same when the bearings are taken, then when you plot the position lines, you will find that the NDB which is further inland gives greater error.
(2) When flying over the water if you take two/three bearings to make a fix, and if errors due to coastal refraction are present, then the fix you make will put you coastward from your true position. Do a little plotting exercise and check it. First plot a fix from three position lines unaffected by coastal refraction; then plot another fix from the position lines in error.
(3) In Fig. 5.19 an aircraft is flying a track of 060° (T) in no wind conditions. When in position A it obtains a relative bearing of 050°R from NDB, C. Now if it waits until the bearing has changed to twice the original value, i.e. 100°R, its position at that time is at B. In this situation, we have a triangle ABC in which side AB equals side BC. This enables the pilot to calculate his distance from NDB, C, when at B. This distance is equal to distance AB which he can calculate from knowledge of his ground speed. (Our mariner forefathers who used this technique for navigation

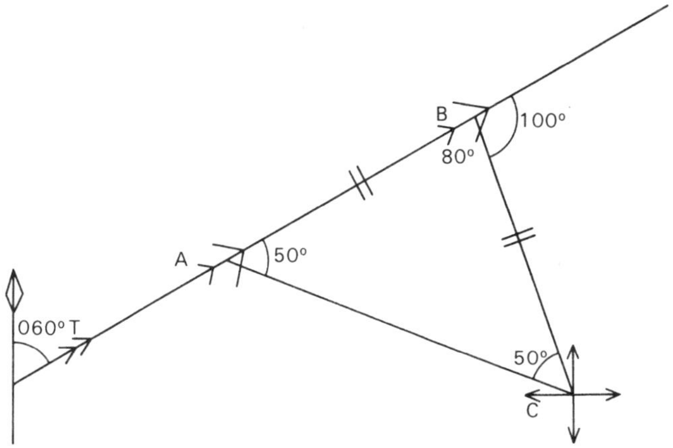

Fig. 5.19 Distance from station.

with visual bearings of headlands, knew it as 'doubling the angle on the bow'.)

When there is wind, allowance must be made for the drift applied in the ADF reading.

(4) If an aircraft flies from one beacon to another and the relative bearing remains fairly steady, the wind velocity may be found on arrival over the second beacon. The time between the two beacons should not be too short. For example, an aircraft passes over beacon A at 10.00 hrs, TAS 180 kt, Hdg(M) 157°. Later the ADF tuned to the NDB, A, gives a steady relative bearing of 186° and the aircraft passes over beacon B, 49 nm from A at 10.20 hrs. Var 8°W.

Aircraft's ground speed: 49 in 20 = 147 kt. Hdg(M) 157, Var 8°W; Hdg(T) 149, Drift 6°S, Tr(T) 155. Now the values of TR, HDG, TAS and GS are put on the computer to give wind velocity (WV) which is 123°/37 kt in our illustration.

Test questions

(1) List the factors that affect the accuracy of ADF indications.

(2) Explain briefly 'night effect'.

(3) List the factors that contribute to restricting useful NDB ranges.

(4) List the factors that affect the field strength of the NDB radiation.

(5) Explain briefly how station interference may affect ADF performance.

(6) What minimum protection is provided when flying within protection range in the UK?

(7) The emission of an NDB is listed as N0N A2A. Explain how you would use BFO before you are ready to take a bearing on it.

(8) Describe the procedure to follow to obtain a QDM from an NDB.

(9) Why is the promulgated range for an NDB valid only during the day time?

(10) What is the guide line to be followed when deciding whether an NDB should have N0N A1A or N0N A2A emission?

(11) NDBs transmit:
(a) vertically polarised signals in the MF band
(b) horizontally polarised signals in the HF band
(c) phase comparable signals in the MF/HF band.

(12) The frequency band(s) chosen for NDBs is/are:
(a) upper LF and lower MF (b) VHF and above (c) HF.

(13) The rated coverage of homing and holding NDBs is a range of approximately:
(a) 25 nm (b) 10 nm (c) 50 nm.

(14) When using ADF for en-route navigation, the bearing obtained is:
(a) magnetic bearing (b) true bearing (c) relative bearing.

(15) The promulgated protection range for an NDB is applicable:
(a) during daytime only (b) during night-time only
(c) throughout the 24 hours, but it is most prone to error around dusk and dawn.

Chapter 6
VHF OMNIDIRECTIONAL RADIO RANGE (VOR) AND DOPPLER (DVOR)

An earlier navigation aid, radio range, operating in the LF/MF band served aviation for a period following the Second World War. It had its limitations inherent with lower frequencies and at best it could produce only four fixed tracks. A need for a more flexible and reliable aid soon became apparent with the expansion of aviation and VHF omnidirectional radio range (VOR) emerged as its successor. It was officially adopted by ICAO in 1960 as a standard short-range navigation aid.

VOR theoretically produces an infinite number of tracks, it is practically free from static and does not suffer from night effect. Consequently it could be used with confidence at any time throughout the 24 hours. The indications are in terms of deviation to the left or right from the selected track. Information may be fed to an RMI to give QDMs. When frequency is paired with distance measuring equipment (DME), range and bearing information provides instantaneous fixes.

Principle of operation

The principle of VOR is bearing measurement by phase comparison. It will be remembered that an NDB transmits an omnidirectional signal and the aircraft's loop aerial converts it to a directional one. A VOR transmitter does this work on the ground and the airborne receiver receives directional information. The ground station transmits two separate signals as follows:

Reference signal

The reference signal is an omnidirectional continuous wave transmission on the station's allocated frequency. It carries a 9960 Hz sub-carrier which is frequency-modulated at 30 Hz. Being an omnidirectional radiation, its polar diagram is a circle. This means that at a given range from the transmitter, the same phase will be detected by an aircraft's receiver on all bearings around it. It will be noticed in Fig. 6.1 that the phase pattern produced is independent of the receiver's bearing from the station. In the receiver, the 30 Hz component of this transmission is used as a reference (or datum) for the purpose of measuring the phase difference.

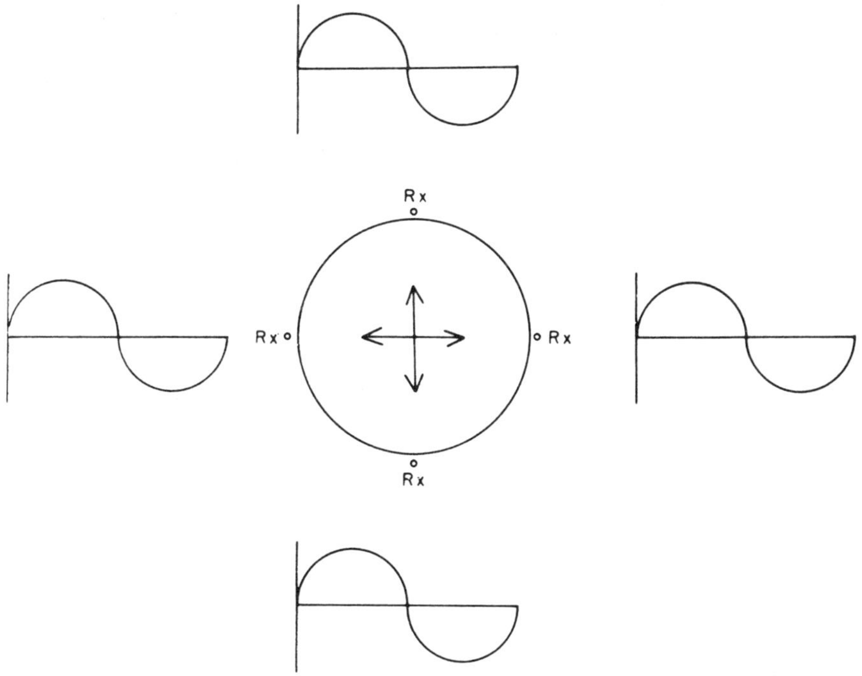

Fig. 6.1 Reference signal – same phase in all directions.

Variable or directional signal

This is again transmitted on the station frequency and the radiated pattern produces a polar diagram of a rotating figure of eight. And by rotating it 30 times per second the signal is given the character of a 30 Hz amplitude modulation. This simply means that the received signal will rise to a maximum and fall to zero value 30 times a second.

Derivation of the phase difference

When the reference signal is combined with the directional signal, a rotating cardioid results. Unlike the cardioid of an ADF loop aerial, the VOR cardioid does not have a null position and in strict terms, a VOR cardioid is called a limacon (see Fig. 6.2). This absence of null is arranged at the transmitter by adjusting the power relationship between the reference and variable signals. The resulting field strength is in the ratios 4:1:4:7 on four cardinal points. In Fig. 6.3 the receivers are shown on four cardinal points, receiving signals of field strength in above ratios, and in Fig. 6.4, when the amplitudes of the incoming signals are plotted on a time axis for the same four positions of the receiver, the signals' directional characteristic is revealed.

Now we have two signals in the receiver; the reference signal producing a

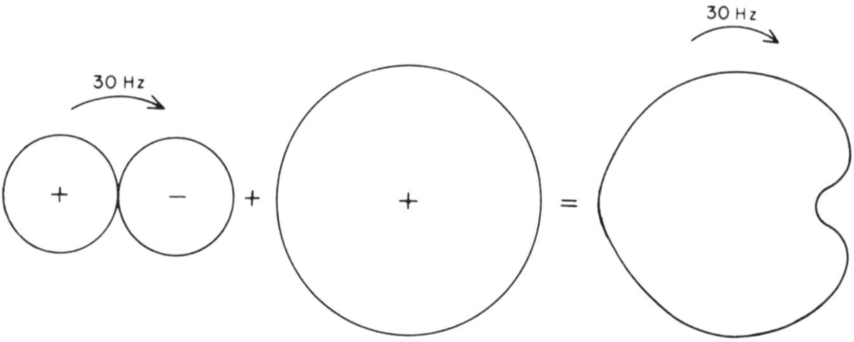

Fig. 6.2 Derivation of limacon.

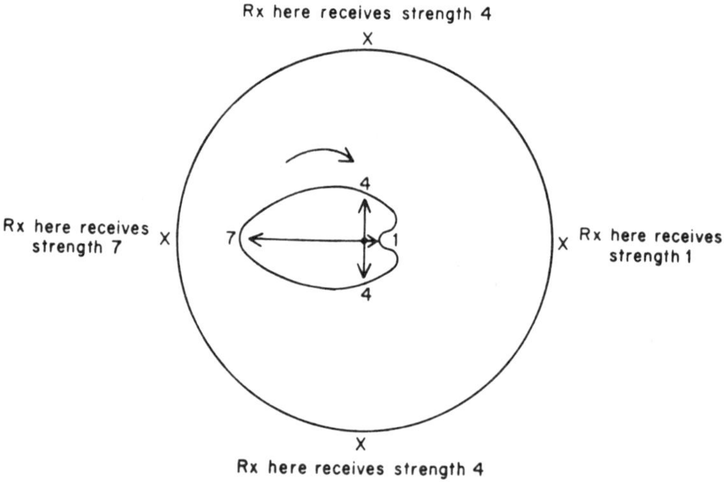

Fig. 6.3 Varying field strengths.

constant phase as shown in Fig. 6.1 and the variable signal giving a bearing-dependent phase, as shown in Fig. 6.4. It will be seen in Fig. 6.5 how the comparison between the two can yield the direction of the receiver. It will also be noted that the receiver north of the station receives both the signals at the same phase, that is, the phase difference between the two is zero. It is deliberately arranged that this zero difference should occur on the station's magnetic north to provide a measuring datum.

It will be observed from Fig. 6.5 that an aircraft on a magnetic bearing of 090 receives a phase difference of 90° and when on 270, it receives a phase difference of 270°. Therefore, when on a bearing of 045, it will receive a phase difference of 45° and when on 227, it will receive a phase difference of 227° and so on. Conversely, when it receives a phase difference of 329°, its magnetic bearing from the station (QDR) is 329, and when it receives a phase difference of 063°, its QDR is 063.

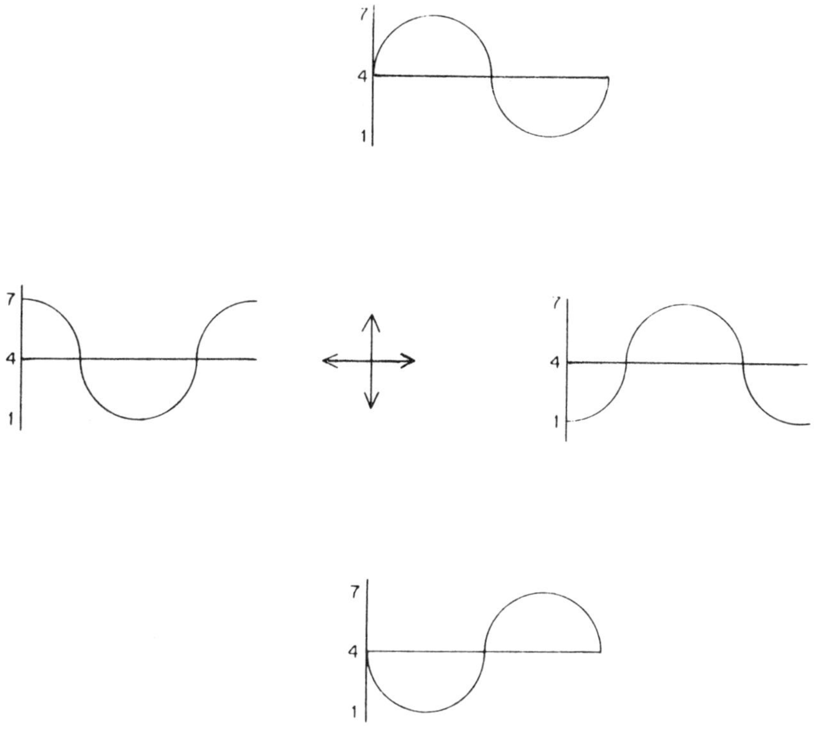

Fig. 6.4 Variable signal phase dependent upon heading.

Thus, a VOR transmitter continuously sends out 360 individual tracks. These tracks are oriented from the station's magnetic north (that is, the station's magnetic variation is applied) and the tracks radiating outward from the transmitter are called 'radials'. It is important to understand the strict meaning of this term so that it is not misused. Study Fig. 6.6.

Airborne equipment

The basic airborne equipment consists of an aerial, a receiver and an indicator.

Aerial. These are small, horizontal dipoles capable of accepting horizontally polarised signals in the frequency band 108 to 118 MHz. They should be installed so that they offer omnidirectional cover to VOR signals but receive no interference from the VHF aerial operating on RTF communication channels. A VOR aerial also accepts ILS localiser signals which are in the same frequency band.

Receiver. A VOR receiver compares the phases of the reference and variable signals and feeds the extracted phase difference in a suitable form to the

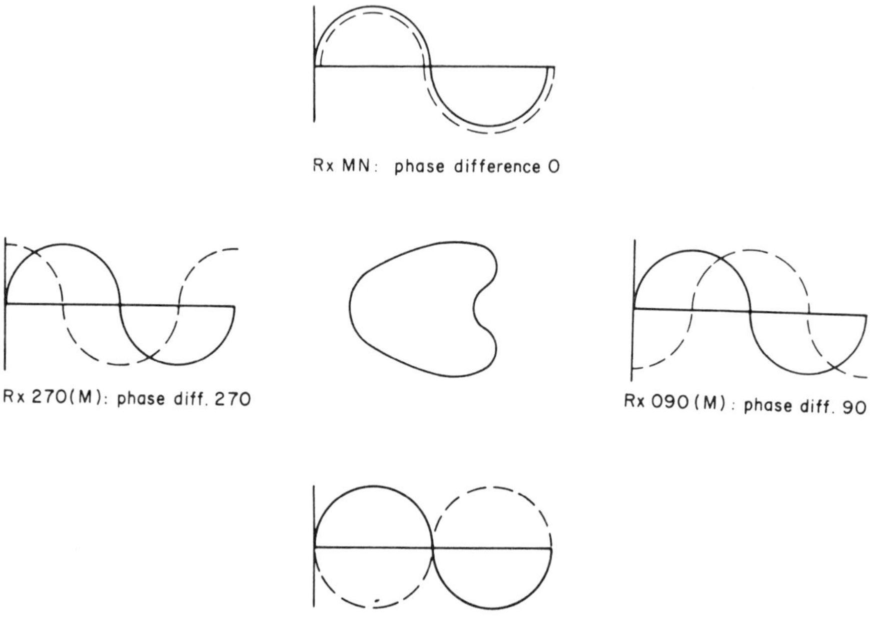

Fig. 6.5 Varying phase difference.

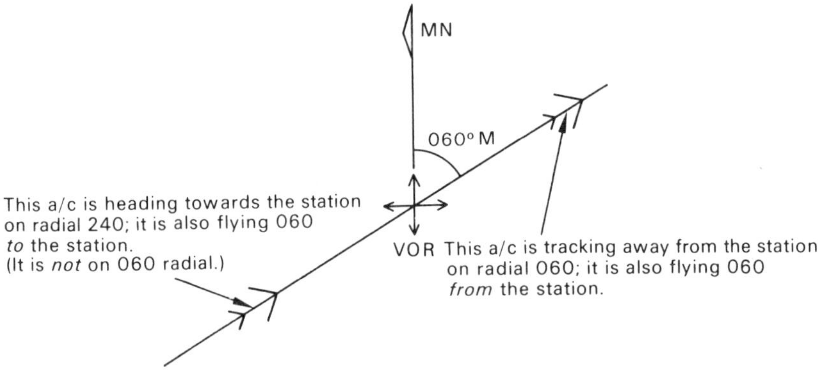

Fig. 6.6 Radials.

various components of the indicator. The two signals are processed through two different channels, and their carriers are filtered out at appropriate stages. A 30 Hz FM reference signal is converted by a discriminator to a 30 Hz AM signal (see Fig. 6.7) and it is then compared in phase with the 30 Hz AM variable signal in the phase detector unit. If the two signals are in phase, the circuits are in balance and the indicator is indicating correct bearing. If the two signals are not in phase an error signal is produced in

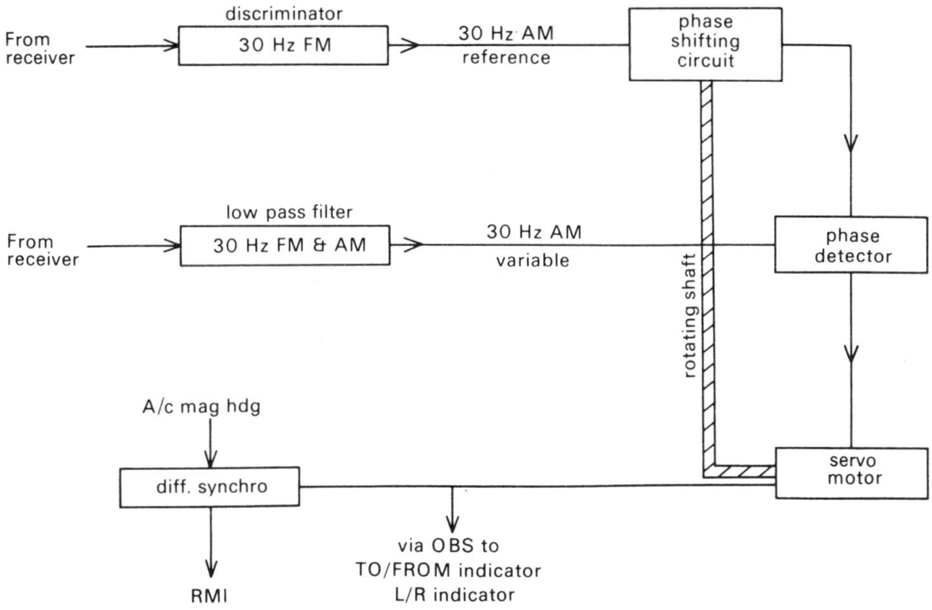

Fig. 6.7 Components of airborne system.

the phase detector which energises the servo motor. The servo motor is connected to a phase shifting circuit via a rotating shaft. On being energised, the servo motor turns the shaft in one or other direction to shift the phase. When the two phases are made alike in the phase detector, the error signal is cancelled and the system comes to rest. The angular rotation of the shaft is the measure of the phase difference.

Indicator. The indicator consists of three basic components which may all be mounted in a single unit or installed separately or in combination. The three basic components are:

- Omni-bearing selector (OBS)
- TO/FROM indicator
- LEFT/RIGHT deviation indicator.

Two different types of indicators are shown in Fig. 6.8. The function and the method of use of these components is explained in the following paragraphs.

(1) ***OBS.*** The OBS control knob is used by the pilot to select the *magnetic track* he wishes to fly to or from a VOR station. To home to the station he may select the track which will take him to the station (or the track he is required to follow) or its reciprocal, which is the radial from the station. For example, a VOR is to be reached on a magnetic track of

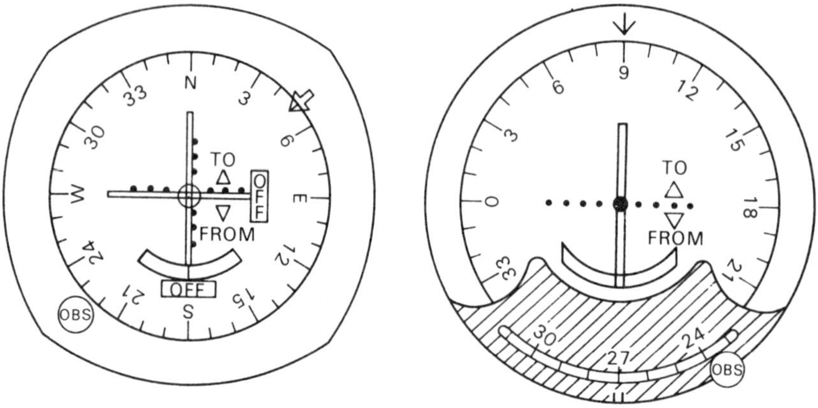

Fig. 6.8 VOR indicators.

050°. The pilot may select 050° or 230° on the OBS. Similarly when flying away from the station on the same track, he may again use either of the above settings. As we will soon see, the selected track affects TO/FROM and L/R indications and so he chooses the best setting.

When the track is selected, the vertical left/right needle is displaced from its central position either to the left or to the right, unless the aircraft is on that selected radial at that time. Alternatively, a pilot may use his VOR just to obtain a bearing, in which case he centralises the needle by use of the OBS control.

(2) ***TO/FROM indicator.*** When the required magnetic track has been selected, the TO/FROM display will indicate either TO or FROM, according to whether the selected radial or its reciprocal is nearer to the aircraft's position. This indication will change when the aircraft crosses a line 90° to the selected track. The rate of change will depend upon the range of the aircraft from the ground station (when not flying overhead). In the following illustration, Fig. 6.9, all aircraft have selected 010 on the OBS.

Aircraft A, B and E will have TO displayed on the indicator since in all three cases if they wish to reach the station on magnetic track of either 010 or 190, then 010 is the nearer. For similar reasoning, aircraft C and D will indicate FROM suggesting that not the selected bearing 010 but its reciprocal 190 is the nearer radial to take him to the station.

There is another way of looking at it. In position A, B or E and having selected 010, ask yourself 'Does this setting take me to the station?' If the answer is yes, TO is indicated. In position C or D the answer is no and FROM is indicated.

(3) ***L/R deviation indicator.*** The indicator consists of a vertical needle which moves left/right across the face from its central position to indicate

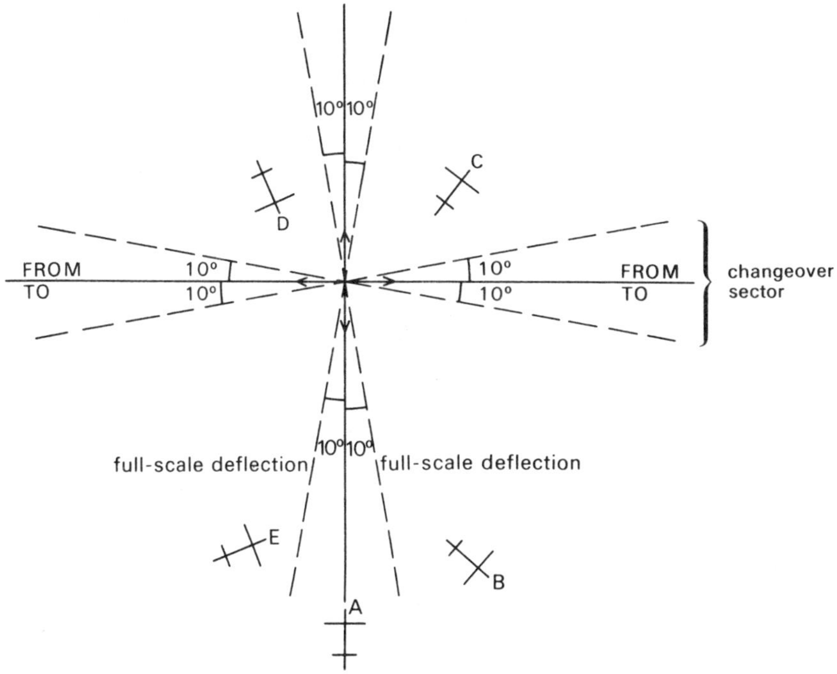

Fig. 6.9 TO/FROM indications.

deviations. The movement is against an angular scale shown by a number of dots. When the aircraft is on the radial the needle stays in its central position. When the aircraft is not on the selected radial, the needle indicates the difference between the selected radial and the radial the aircraft is actually on by moving out to left or right. The amount of deflection is the measure of the angular distance to the selected radial and is estimated from the dots scale.

A full-scale deflection occurs when the aircraft is 10° or more away from the selected radial. This means that no movement of the needle takes place from its maximum-deflection position until the aircraft is within 10° of the selected radial. The instrument may have either a four-dot scale or a five-dot scale. If it is a four-dot instrument then one dot deflection indicates a deviation of $2\frac{1}{2}$°; similarly, on a five-dot instrument it will equal 2°.

Whether you would follow the needle by steering in the direction of the displaced needle or you would go against the needle depends on your physical position with regard to the selected radial, and is independent of the aircraft heading. However, since we must steer the aircraft to gain the radial we must translate this indication in terms of our present heading. The rule is: follow the needle to regain the radial if your heading and the selected bearing are in general agreement.

This rule immediately makes it clear that when homing to/from a station if the track set on OBS is the same as the track we want to follow then the L/R indications will be correct. When the aircraft deviates, say, to right of the track, the needle will move to the left, indicating that the pilot should turn to the left. To make this clear, say you are flying to a VOR on track 090 and you are going to continue on that track past the VOR, select 090 and not 270.

Going back to Fig. 6.9 with this in mind:

Aircraft A will have its needle central.

Aircraft B will have a left turn indication as its heading is virtually the reciprocal of the selected radial.

Aircraft C has 010 selected and its heading is in general agreement and therefore its indication is correct, that is, it should turn left to intercept the radial.

Aircraft D will have an incorrect indication of a right turn.

Aircraft E will have a correct turn right indication.

Now study Fig 6.10 for further familiarisation with indications.

Failure warning flag. All indicators employ a device to warn the pilot when the system has failed. An OFF flag indicating a failure will appear on the face of the indicator in the event of any of the following occurring:

- failure of the aircraft's receiving equipment
- failure of the ground station equipment
- failure of the indicator, or
- where the signals being received are weak or the aircraft is out of range.

VOR frequencies

VOR operates in 108 to 117.95 MHz band as follows. (A9W emission.)

Band 108–112 MHz. This is primarily an ILS frequency band but ICAO prescribes that it may be shared with VOR if it is not fully subscribed. Thus, normally this band is shared between ILS localiser and short range (terminal) VORs (TVORs). VOR uses frequencies on 'even' first decimals (108.20, 108.25, etc.) and ILS uses 'odd' first decimals (108.10, 108.15, etc.).

Band 112–117.95. VOR (odd and even decimals).

ICAO's recommendation that the frequency spacing should be reduced to 50 kHz has been approved by the CAA and the future allocations will be based on this spacing. We thus have VORs operating on frequencies, for example, 112.30, 112.35 and so on.

Use of VOR

VOR, like ADF, may be used in a variety of ways; it is more reliable and certainly easier to use once the principles are understood. You may home to your destination aerodrome from any direction. You may fly cross-country

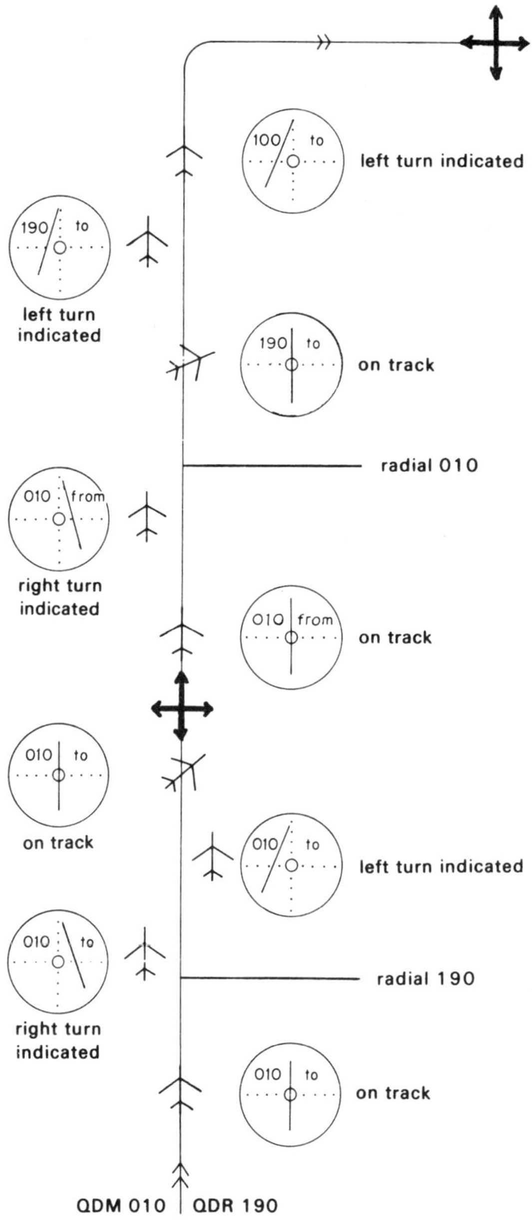

Fig. 6.10 Indicator readings.

tracks from beacon to beacon, as on an airway. If your route does not take you over VOR stations, you may still make use of them in your navigation. You may carry out holding procedures and make an approach to let-down at your destination.

Before you actually use the VOR in the air for any of the above purposes it is important that you go through the following routine.

(1) First check that you are within the designated operational coverage (DOC) of that station. This information is given in the communications section of the UKAIP. Outside the UK you may find this information listed under 'range and altitude' or 'protection range and altitude'. If you are not within the quoted DOC of the station which applies both day and night, you must not attempt to obtain the information even if the display units seem to give reliable indications. This is because the signals are not protected from interference from other beacons.
(2) Having satisfied yourself that you are within the coverage, switch on the equipment. It may take a couple of minutes to warm up.
(3) Select the station frequency, and if the indications are required on RMI, select VOR on the green pointer.
(4) Check that the warning flag clears out of the window.
(5) Identify the station.

You are now ready to use your VOR. Let us discuss the procedures.

Homing on VOR

It is possible that you may be able to home to a station from any direction, that is, on any radial. In that case, as soon as you are receiving satisfactory signals you are on a VOR radial and ready to proceed.

But it is more likely that your approach direction is restricted for one reason or the other, e.g. a danger area in the way or presence of high ground and so forth. ATC imposes movement restriction to ensure a smooth traffic flow by prescribing inbound and outbound routes. These routes are defined by VOR radials and to home to the station you must use one of the routes most conveniently placed. Well, how do we get on this radial? With VOR it is relatively simple. In Fig. 6.11, the aircraft in position 1 wishes to join radial 090 from VOR A. The inbound track is 270°(M). The pilot has some idea of his position in relation to the required radial. If he is not sure, his QDM to the station is quickly checked by turning the OBS control until the needle is central and TO is indicated. The following procedure is adopted.

Set your track, 270 on the OBS counters. TO/FROM display will indicate TO (as this is your track to the station) and the L/R needle will swing to maximum deflection, indicating a right turn in our illustration. The indication of right turn is ignored as your own heading is not near 270°. With the knowledge of your position with regard to the selected radial, set a heading that will give you a comfortable angle of interception.

When in position 2, you are 10° away from the selected radial and the L/R needle will show an inward movement. Assess your angle of lead, taking into account your radius of turn, the angle through which you have to turn and the distance from the station. The movement of the L/R needle may give you an indication of how fast or slow you are approaching your radial. The further the distance from the station, the slower its inward movement.

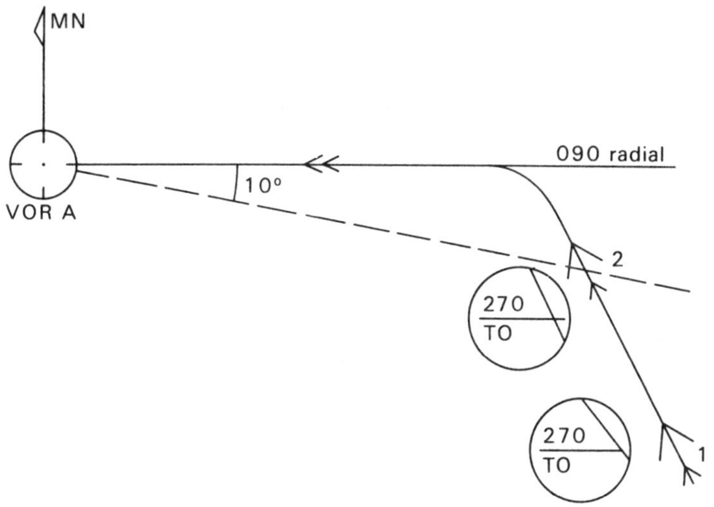

Fig. 6.11 Homing procedure.

If you have assessed your angle of lead correctly, on completion of the turn you should be on the radial, the L/R needle in the centre position and the TO/FROM indicator indicating TO. You are homing to the station.

Apply drift to fly 270 track. If the drift is correct, the needle will remain central. If the needle starts to ease to one or other side, follow the needle. One or two alterations in 5°/10° steps will give you the correct heading. Once on the correct heading, just monitor the needle for any indications of a wind change. The aircraft will arrive overhead VOR by keeping the needle central.

If the flight continues on the same track past the VOR beacon, the OBS setting remains unaltered. The TO/FROM indication will change to indicate FROM, and you follow the needle to correct for any deviations.

Taking a bearing

- Select a station within DOC
- Switch on
- Check the warning flag
- Identify the station
- Turn the OBS control until the needle is central
- Read off the radial indicated on the OBS counters and note the time
- Check the TO/FROM indication. If TO is indicated, it is QDM; if FROM is indicated, it is QDR.

Station passage and cone of confusion

As the station is approached, radials get closer to each other and the needle becomes increasingly sensitive. Minor deviations are shown up calling for small adjustments of heading to keep the aircraft on the radial. When closer, the needle oscillates hard from side to side, the OFF flag may momentarily appear and TO/FROM display swings between TO and FROM, not being sure whether it is coming or going!

The cause of these erratic indications is the presence of what is called the 'cone of confusion' overhead the beacon.

The propagation specifications, as recommended by ICAO, require the signals to be transmitted up to 40° in elevation. In practice, the modern equipment is capable of radiating signals of up to 60° to 80° above the horizon. But it still leaves a gap overhead, in the form of a cone where no planned radiation takes place. While passing through this zone the receiver comes under influence of weak overspill, causing confusion to the indicators.

Once through the cone all indications settle down to indicate correctly, TO/FROM indication changing over to FROM. The time you will remain in the cone depends on your height and the ground speed.

Identification

In the UK, VORs transmit a 3-letter aural morse group at a rate of approximately seven words per minute, at least once every ten seconds. This is the ICAO recommendation. Ident may also be given in speech form, e.g. 'This is Miami Omni Range', immediately followed by ident in morse. The voice channel may also be used to pass significant weather information to the aircraft.

Monitoring

All VOR ground stations' transmissions are monitored by an automatic monitor located in the radiation field near the station. The monitor will warn the control point and either remove the identification and navigation components from the carrier or switch off the radiation altogether in the event of any of the following circumstances:

(1) a change in bearing information at the monitor site in excess of 1°
(2) a reduction of more than 15% in the signal strength of the two 30 Hz modulation components (or either one of them) or the RF carrier itself
(3) failure of the monitor itself.

When the main transmitter is switched off a standby transmitter is brought into operation. This takes a certain amount of time before its radiation stabilises. During this period the bearing information radiated can be incorrect, and as a warning to the users, no ident signals are transmitted until the changeover is complete.

When an approach to an airfield is being made using a short-range TVOR, it is vital that the pilot continuously monitors the ident signals in order to ensure that the bearing information being received is correct and the approach being made is safe.

Factors affecting VOR ranges

(1) ***Transmission power.*** The higher the power, the greater the range subject to altitude. En-route VORs with a power output of 200 watts achieve ranges of around 200 nm. TVORs normally transmit at 50 watts.

(2) ***Transmitter and aircraft height.*** Because VOR transmissions are in the VHF frequency band, the theoretical maximum range depends on line-of-sight distance (in practice, slightly better due to atmospheric refraction). For calculating theoretical ranges for various heights, the VHF formula given earlier is used. This is repeated here for convenience.

$$\text{Max range (in nm)} = 1.25\sqrt{H_T} + 1.25\sqrt{H_R}$$

where H_T is the height of the transmitter amsl, and H_R is the height of the receiver amsl.

The three factors involved in VHF formula are range, transmitter height (amsl) and the aircraft height (amsl). Knowing any two of these, the value of the third factor can be found.

Examples

(a) If the transmitter's altitude is 100 ft and the aircraft's altitude 12 500 ft, at what maximum range would the aircraft receive the VOR signals?

$$\begin{aligned} \text{Range} &= 1.25\sqrt{H_T} + 1.25\sqrt{H_R} \\ &= 1.25\sqrt{(100)} + 1.25\sqrt{(12\,500)} \\ &= (1.25 \times 10) + (1.25 \times 112) \\ &= 12.5 + 140 \\ &= 152\,\text{nm (near enough)} \end{aligned}$$

(b) At what altitude should an aircraft be to receive VOR signals at a range of 130 nm if the VOR transmitter is 500 ft amsl?

$$\begin{aligned} \text{Range} &= 1.25\sqrt{H_T} + 1.25\sqrt{H_R} \\ 130 &= 1.25\sqrt{(500)} + 1.25\sqrt{H_R} \\ 130 &= 1.25(\sqrt{(500)} + \sqrt{H_R}) \\ \frac{130}{1.25} &= \sqrt{(500)} + \sqrt{H_R} \\ 104 &= 22 + \sqrt{H_R} \\ \sqrt{H_R} &= 104 - 22 = 82 \\ H_R &= 82^2 \\ H_R &= 6725\,\text{ft} \end{aligned}$$

(3) ***DOC: Protection range and altitude.*** At present there are only 20 channels in the 108–112 MHz band and a further 60 channels in 112–117.95 MHz band. Thus, the VOR band is very limited in channel space and like NDBs it is necessary to protect the wanted signals from interference due

to unwanted signals. The protection is given not only in range but in altitude as well. Protection in altitude is possible because the signals travel in straight lines.

This need for protection becomes apparent when one looks at the problem from the planner's point of view. Suppose it is required to establish a VOR to give interference-free reception at 25 000 ft. First the distance the wave must travel to gain that altitude due to the earth's curvature is calculated; the Tx is at sea level (Fig. 6.12).

$$\begin{aligned}\text{Range} &= 1.25\sqrt{(25\,000)}\\ &= 1.25 \times 158\\ &= 198\,\text{nm}\end{aligned}$$

Based on this calculation, another station, operating on the same frequency channel must be at least another 198 nm further away so as not to produce interfering radiation. This gives a geographical separation distance of 396 nm.

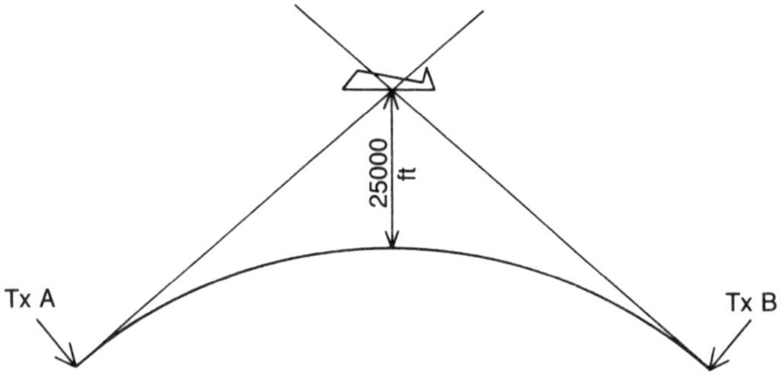

Fig. 6.12 Separating VORs.

But this figure is derived from consideration of only one factor, that is, the line-of-sight propagation. There are other calculable factors which must be considered, e.g. the relative radiation strength of the two transmitters, attenuation of the signals due to range and altitude and the degree of protection required. The planners normally work on providing a protection of 20 dB. At a fair estimate these factors would add another 100 nm. Thus, to establish this VOR, the planners must ensure that no other VOR within a radius of 500 nm transmits on the same channel.

In a practical approach, a transmission is protected only to the extent necessary, in range and altitude, and not necessarily to the maximum line-of-sight range. Further, by increasing the radiation power of one transmitter, the separation distance from other co-channel transmitters may be reduced. Fig. 6.13 illustrates the practical scheme of providing protection. In the UK these values are published under the term designated operational coverage (DOC).

In Fig. 6.13 let us say that the protection range and altitude for VOR 1 is

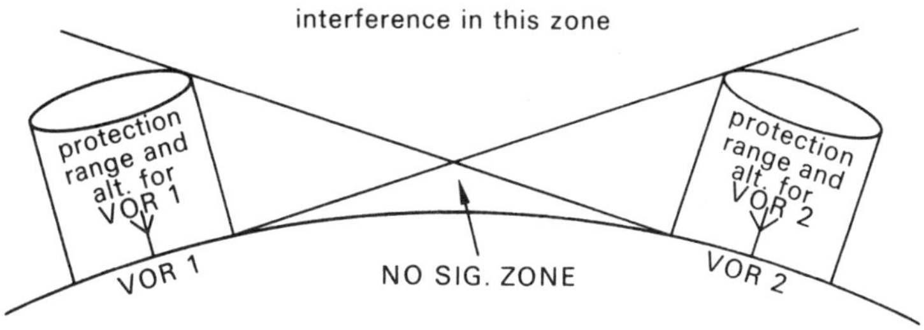

Fig. 6.13 Protection range and altitude.

given as 50/25 000. An aircraft at a range of 50 nm from the transmitter is protected from interference to an altitude of 25 000 ft. Another aircraft, say at 30 000 ft will pass over the protected volume of airspace, at less than 50 nm (observe the diagram) and it has a fair chance of receiving interference-free signals. These should not be used. Both altitude and range limits must be observed.

(4) ***Nature of terrain.*** Uneven terrain, intervening high ground, mountains, valleys, man-made obstacles: all these features affect VOR propagation. The signals are screened, shadowed, reflected, bent, split and so forth, giving erroneous information. Where such effects have been known to exist, sectors giving errors are marked out and noted in the UKAIP. Typical information reads: 'errors up to $5\frac{1}{2}°$ may be experienced in sector radials 315° to 345°'. You may also find VORs in respect of which more than one protection altitude and range are prescribed.

Factors affecting accuracy

(1) ***Site error.*** Irregular or uneven terrain, physical obstacles, etc. in the vicinity of a VOR transmitter also affect its directional propagation. Stringent requirements are laid down by ICAO regarding site contours, presence of structures, trees, wire fences, etc. Even the overgrowth of grass affects the signals. Error introduced in radiated directions by these features is called VOR course displacement error. As we learnt earlier, VORs are ground-monitored to an accuracy of ±1°.

(2) ***Propagation error.*** The signals having left the transmitter giving an accuracy of ±1°, suffer further inaccuracies as they travel forward. Uneven terrain and other features which affect the signals at the site, continue to affect them throughout their passage to the receiver.

(3) ***Airborne equipment error.*** Airborne equipment is required to translate one degree of phase difference to one degree of change of direction. Inaccuracy is introduced in the process to which the indicator makes its contribution when the signals finally arrive for the display.

The above three errors in combination are called VOR aggregate errors; errors listed under (1) and (2) being due to ground propagation and those under (3) due to airborne equipment.

These errors are easily calculable by taking the square root of the sum of the squares of the two types of errors. For example, if the error due to ground propagation is 3.5°, and the error in the airborne equipment is 3°, then

$$\begin{aligned} \text{aggregate error} &= \sqrt{(3.5^2 + 3^2)} \\ &= \sqrt{(12.25 + 9)} \\ &= \sqrt{(21.25)} \\ &= 4.6^\circ \end{aligned}$$

In addition to the above we must consider a further two causes of errors as follows.

(4) ***Pilotage error.*** As the VOR station is approached, the signal strength increases rapidly and the radials get closer. The needle becomes sensitive to minor deviations and the pilot cannot or may not keep his aircraft precisely on the radial. However, because at this stage the radials are very close to each other the lateral displacement of the aircraft from its intended track is small. In planning calculations, this error is given a fixed value of ±2.5°.

(5) ***Interference error.*** This is an avoidable error which affects the indications when using a VOR outside the DOC or when below the line-of-sight altitude. When below the line of sight, if the signals are being received it is obvious that they must be weak signals arriving in the receiver due to reflections and other scatter effects.

The overall accuracy of the information displayed is ±5°; in the worst case due to other random variable errors this may deteriorate to $\pm 7\frac{1}{2}^\circ$.

On the basis of a worst accuracy of $\pm 7\frac{1}{2}^\circ$ on an airway where the navigational information error should be limited to ±5° to $\pm 5\frac{1}{2}^\circ$ (to keep the aircraft within the airway limits), two VORs should not be further apart than 80 nm, as can be seen in the following calculation using the 1 in 60 rule.

$$\begin{aligned} \text{Track error} &= \frac{60 \times \text{dist off}}{\text{dist to go}} \\ 7.5 &= \frac{60 \times 5}{\text{dist to go}} \\ \text{dist to go} &= \frac{60 \times 5}{7.5} \\ &= \frac{300}{7.5} \\ &= 40\,\text{nm} \end{aligned}$$

Thus a 5° error occurs at a distance of 40 nm from the transmitter, setting the limit to its use.

Test VORs

These are installed at certain aerodromes to enable the pilots to test the airborne VOR equipment during preflight checks. The transmitters are called VOTs and the frequencies are published in the States' AIPs. To test the airborne equipment from any position on the aerodrome, just tune in to the channel and centralise the needle. OBS counters should indicate 000 FROM or 180 TO. If they do not indicate within ±4°, the equipment requires servicing.

Advantages of VOR as a navigational aid

- In comparison with the system it replaced, VOR gives its indications in a form which is easy to see and follow
- In theory it provides an infinite number of tracks
- It is free from night effect and practically free from static
- Being a VHF aid its ranges can be accurately forecast before the beacons are sited, thus avoiding interference
- Its left/right deviation indicator can also display ILS signals
- It can be frequency-paired with DME to give fixes
- It incorporates an equipment-failure warning device
- Its channel spacing is much better than NDBs
- Being in the VHF band, its aerials are smaller.

Disadvantages of VOR as a navigational aid

- Its left/right indications do not point to the beacon; thus, for a continuous indication of QDMs an RMI must be used
- Only position lines are available
- High ground and man-made obstructions can cut off, reflect and attenuate the signals
- Numerous beacons are required to give a large area coverage
- It gives line-of-sight ranges only.

Doppler VORs

Doppler VORs (DVORs) are the second generation VORs, the main aim being to improve the accuracy of the signals.

Conventional transmitters suffer from reflections from objects in the vicinity of the site. It was found that the errors due to this could be reduced if the horizontal dimensions of the aerial system were increased. However, this could not be achieved with the conventional method of transmission and a new approach was necessary.

In the DVOR system the reference (or constant phase) signal is transmitted from a central aerial and it is amplitude-modulated. The variable signal is transmitted from a system of about 50 aerials encircling the central aerial and it is frequency-modulated (Fig. 6.14). Thus the modulations are employed in reverse roles. The circle of aerials is 44 ft in diameter to give the necessary

Fig. 6.14 Doppler VOR station.

Doppler shift, compared with the 10 ft high 6 ft diameter dustbin-like structure of earlier conventional VORs. The resultant propagation is much less sensitive to obstructions in the vicinity, i.e. site error is less. The Doppler principle involved is explained in chapter 16.

However, the transmission frequencies are the same, the same airborne equipment can receive and process the signals and, as far as the operation on the flight deck is concerned, there is no difference between VORs and DVORs. Having said that, the CAA has found it necessary to issue a pink Aeronautical Information Circular (AIC) alerting pilots of light aircraft to reports it has received of improper DVOR operation. The improper operation can be in the form of either:

(1) centred pointer with warning flag showing
(2) large bearing errors
(3) very low pointer deflection sensitivity
(4) reversed TO/FROM flag.

(2), (3) and (4) may occur together and may or may not be accompanied by the warning flag, and all of them may occur when flying within the DVOR's coverage area. The AIC alerts pilots to the fact that the effects may be encountered in Europe and identifies a particular UK station. Pilots with the types of receivers/converters fitted in their light aircraft specified in the appendix to the AIC are advised to have the equipment tested by the manufacturers for alternating sideband (ASB) DVOR compatibility. They are also advised to take every opportunity to monitor the operation of their aircraft's installation with ASB DVORs.

Exercises on use of VOR indicators and RBI

(1) An aircraft's VOR is tuned to station A. When 075 is selected on OBS the needle becomes central and the TO/FROM indicator indicates TO. At the same time, an NDB located on the VOR site gives a reading of 012° on ADF. What is the aircraft's heading, and what drift is being experienced?

Answer: The aircraft's magnetic track is 075, and as the ADF indicates 012, the aircraft's nose is offset by 12° to the left. Thus, its heading is (075 − 12) = 063 and the drift is 12° starboard.

(2) An aircraft's heading is 050°(T), and is tracking on a VOR radial, with 068 set on the OBS and TO/FROM indicator indicating TO. The variation is 10°W. What should the aircraft's ADF read from an NDB sited at the VOR station?

Answer: 008°.

(3) An aircraft bears 230°T, distance 15 nm from a VOR station. Local variation is 10°W.

(a) What selected bearing should make the L/R needle central?

Answer: The aircraft's magnetic bearing from the station is (230 + 10W) = 240. Thus, the needle will be central when 240 is selected; it will also be central when its reciprocal, 060 is selected. Thus, the answer is: 240 FROM or 060 TO.

(b) Would the L/R needle indicate left or right, if the bearing selected is 055?

Answer: Turn right (if you draw a simple sketch you will notice that the radial 055 is to the right of the radial 060 when extended to the opposite side of the transmitter).

(4) If an aircraft is on a true bearing of 216° from a VOR station, what is the phase difference between the reference and variable signals arriving in the aircraft's receiver? The variation is 10°E.

Answer: 206°.

(5) An aircraft on a heading of 150°(M) tunes a VOR station, and selects 170 on the OBS. At that time the TO/FROM indicator reads TO and the L/R needle is displaced very close to the maximum deflection position, indicating a right-hand turn. What is the aircraft's approximate position in relation to the transmitter?

Answer: Approximately to the magnetic north of the transmitter.

Chapter 7
RADIO MAGNETIC INDICATOR (RMI)

RMI, as the name suggests, is only an indicator and not an independent navigational aid. It accepts relative bearings from the ADF receiver and phase differences from the VOR to indicate QDMs in both cases. It employs a rotating scale card calibrated in number of degrees and positioned by the aircraft's remote indicating compass. Thus, on the indicator at the 12 o'clock position the aircraft's magnetic heading is read off against a heading index. QDMs are indicated by two concentric pointers, differently shaped, each of which may be energised simultaneously by two like or unlike aids. Fig. 7.1

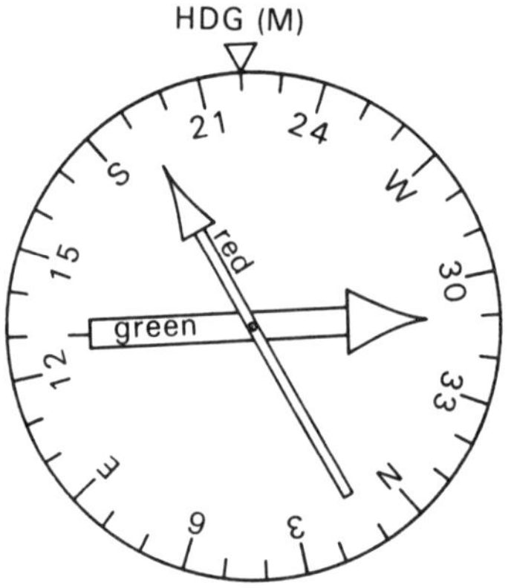

Fig. 7.1 Radio magnetic indicator.

shows a typical RMI. The thin pointer is coloured red and the wide pointer green. By convention the red pointer is called number one needle and the green pointer number two. And again, by convention, number one needle is used for ADF, and number two for VOR indications. In the extremely unlikely event of an errant compass card, only relative bearings can be deduced from the RMI.

RBI relative bearings

In the following illustration, Fig. 7.2, an aircraft on heading 030°(M) has tuned to NDB, X. The ADF gives a relative bearing of 090°R. The measuring

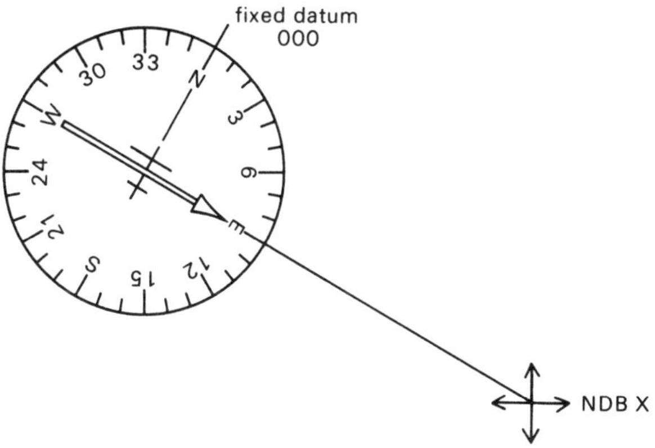

Fig. 7.2 Relative bearing 090°.

datum on the indicator is 000, that is, the fore-and-aft axis of the aircraft. From Fig. 7.2 we can see that the QDM to the NDB is 120° and thus, the relationship between a relative bearing and a QDM is the aircraft's heading. Or,

Rel Brg	090
+Hdg(M)	030
QDM	120

In an RMI, this addition of the heading to the relative bearing is automatically carried out by adjusting the measuring datum. By measuring the pointer's indications from the aircraft's magnetic heading for datum (Fig. 7.3), the aircraft's QDM is read off. It will be noted that as the aircraft

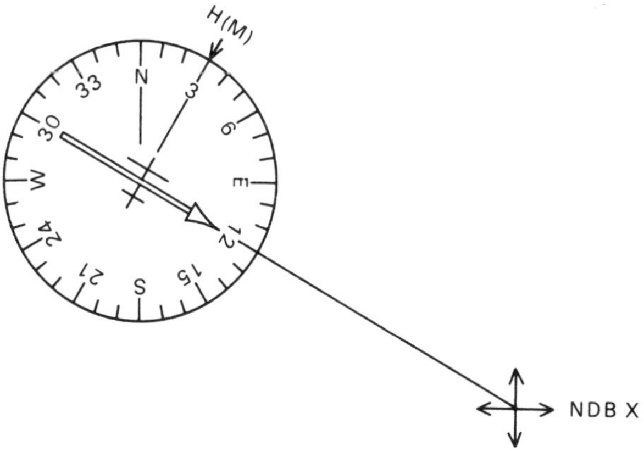

Fig. 7.3 QDM assessment.

measures ADF relative directions with reference to its magnetic heading, when this QDM is converted to a true bearing for plotting from the station, the aircraft's magnetic variation (*not* the station's) must be applied.

VOR phases

VOR airborne equipment receives phases of two different signals (reference and variable) and derives QDR by taking the difference between the two. Its reciprocal is QDM. An RMI accepts relative bearings and not QDMs. Therefore the VOR phase difference giving QDM is first converted to give relative bearing. This is achieved by use of a differential synchro in the VOR navigation unit which subtracts the aircraft's magnetic heading to give relative bearing before the information is fed to the RMI. The RMI subsequently adds the aircraft's heading to give the QDM. In the illustration Fig. 7.4, the aircraft is heading 030°(M) and the RMI indicates 120° QDM from VOR, Y. The magnetic variations at the VOR station and the aircraft position are different. The process of indication is as follows.

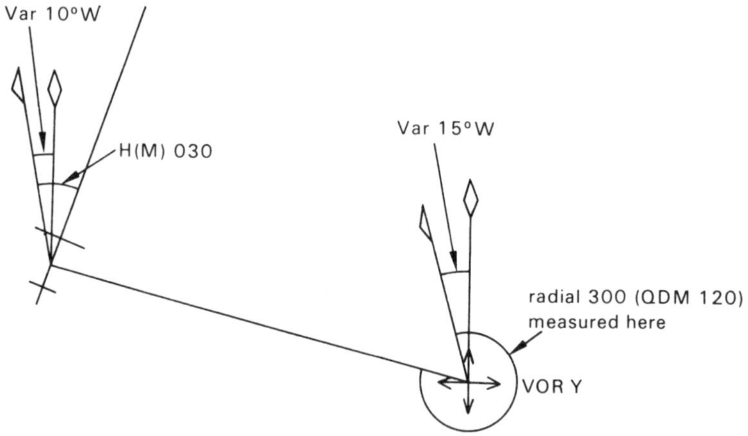

Fig. 7.4 VOR QDM.

QDM received from VOR	120	
−Aircraft H(M)	030	aircraft variation
Rel Brg	090	
+Aircraft H(M)	030	RMI card, aircraft variation
Indicated QDM	120	Station variation

Thus, first the magnetic heading is subtracted, and then the same value is added. These two operations mutually cancel each other out and the resultant indication is affected only by the variation at the VOR station. Therefore, when converting QDM to true values for plotting, variation at the station position is applied.

For reasons of mutual cancellation, the use of the aircraft's heading introduces no error in the indication of VOR QDM. What would be in error is the calculation as above to the stage where the relative bearing is found. This is because the aircraft is using a variation which is different from the station variation. We will look into this in detail.

Discrepancies in indications

(1) If the aircraft's variation is different from the station's variation, the indication of the relative bearing is incorrect but the QDM indicated is still correct.

In Fig. 7.5, aircraft A and B are on a magnetic heading of 030°. The variation at A's position is 20°W, the variation at both B's and the station's position is 10°W. Both aircraft are on a QDM of 100° from VOR, X. From the figure it will be apparent that the relative bearing of the VOR transmitter from aircraft A is 080°R and from aircraft B, 070°R. Both aircraft calculate a QDM of 100° as follows.

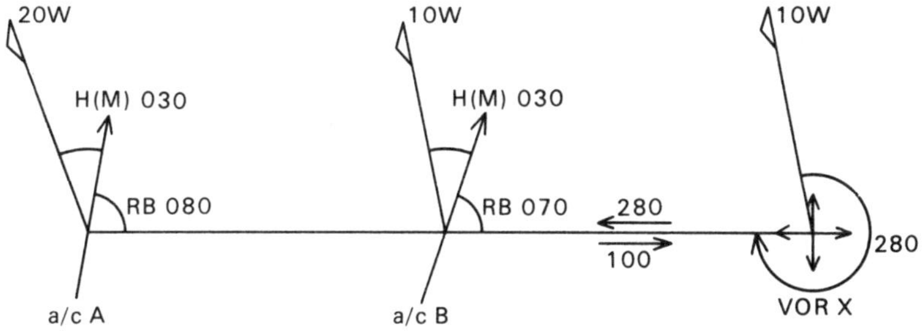

Fig. 7.5 Variation effect.

Aircraft A:	QDM	100	
	−Hdg(M)	030	
	Rel Brg	070	This is what will be read off against a fixed RB scale; its true relative bearing is 080°
	+Hdg(M)	030	
	QDM	100	
Aircraft B:	QDM	100	
	−Hdg(M)	030	
	Rel Brg	070	
	+Hdg(M)	030	Indicates correct relative bearing (same variation)
	QDM	100	

(2) If an NDB is located on the VOR site, QDMs from the VOR and the NDB as displayed by an RMI will be different, if the variation between the aircraft and the station positions is different.

In Fig. 7.5, the VOR gives a QDM of 100°. This is measured at the VOR site. An NDB on the same site would give a relative bearing of 080°R to the aircraft, using local variation. An RMI adds relative bearing to the heading to give the QDM. Thus the QDM displayed is (080° + 030°) = 110°. Simultaneously, the VOR pointer will indicate a QDM of 100°.

Similar discrepancies in the indications will be observed due to convergency between aircraft and the station positions. This is because, like variation effect, the aircraft's true north (from which it derives its magnetic heading) is different from the true north direction of the station. When QDM from a VOR is displayed, the relative bearing indication will be approximate and when QDMs from VOR/NDB are displayed, the two QDMs will show a difference. The discrepancy will be the value of convergency.

Thus, variation and convergency affect the indications as explained above. However, these considerations are more of academic nature than practical. In practical use of the equipment, an aircraft not too far from the transmitters will experience only minor discrepancies and these may be disregarded. And the QDMs from VORs are always shown correct.

However, if these two QDMs are required to be reduced to plotting values, the procedure should be as follows:

QDM derived from an NDB: apply variation at the aircraft and plot the reciprocal; QDM derived from a VOR, apply station variation and plot the reciprocal. For fuller explanations on plotting techniques, the student should consult *Ground Studies for Pilots* volume 2.

Advantages of RMI

- QDM/QDR are indicated continuously and read off directly. The tail end of the pointers indicate QDRs
- Using two beacons, instantaneous fixes are obtained
- RMI indications provide a very useful guide when initially joining a radial for VOR homing
- The indicator itself can be used for homing
- Magnetic headings can be read off together with QDMs
- Approximate relative bearings may be read off against a fixed RBI scale or assessed visually.

Homing to/from a station

Fig. 7.6 shows the indications on RMI and RBI and Fig. 7.7 shows indications on the VOR L/R deviation indicator and RMI when homing to/from a station.

VOR – NDB – RMI exercises

(1) An aircraft bears 220°(T) distance 20 nm from a VOR beacon. Its heading is 055°(M) and variation is 20°W.

(a) What selected bearing should make the L/R needle central?

(b) Would the L/R needle indicate turn to left or right if the selected radial was 055?

(a) Brg Mag = 220°(T) + 20°W (Var)
= 240
∴ QDM = 240 − 180
= 060 which, when selected will make L/R needle central.

Also, its reciprocal, 240 will make the needle central.

∴ 060 TO and 240 FROM

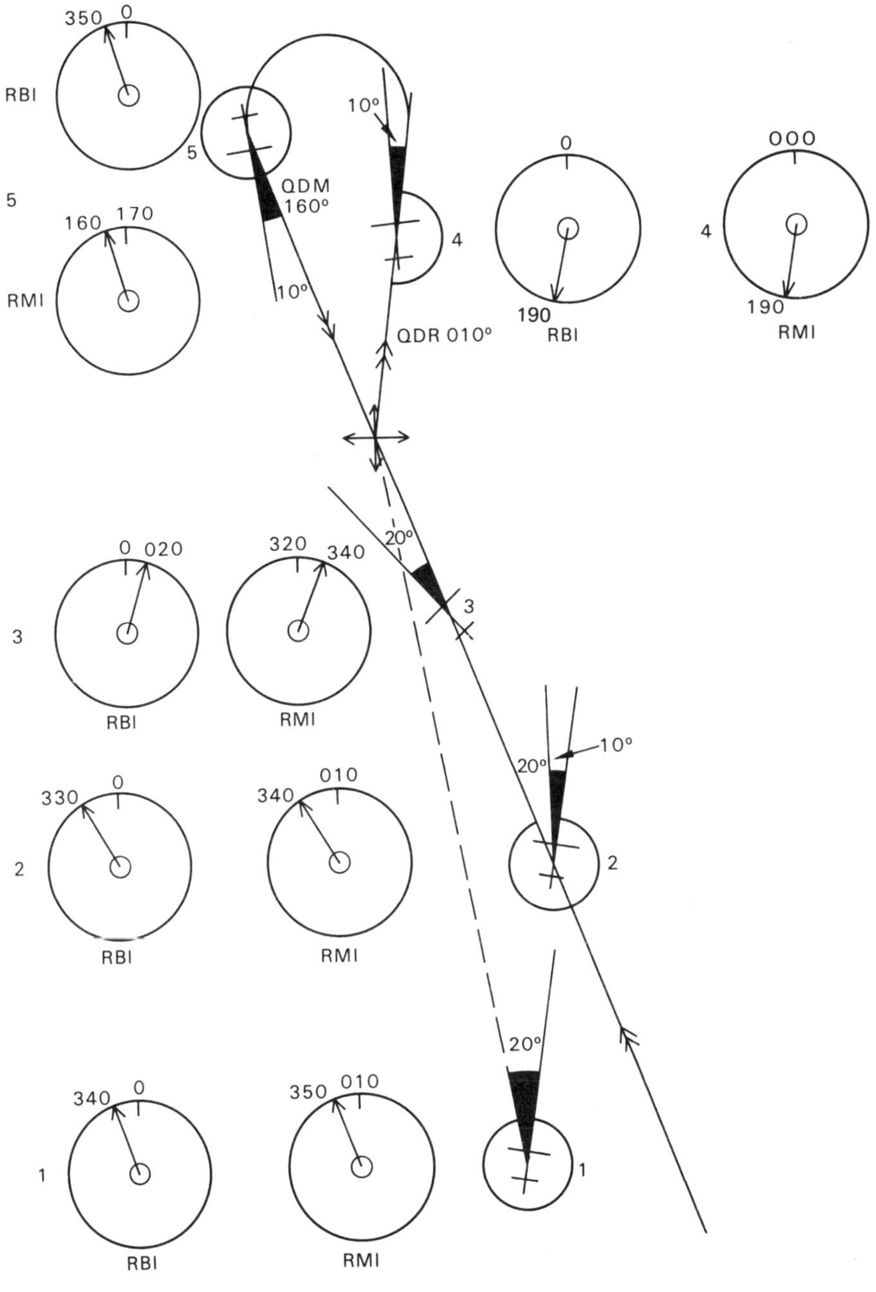

Fig. 7.6 Homing procedures using RBI and RMI.

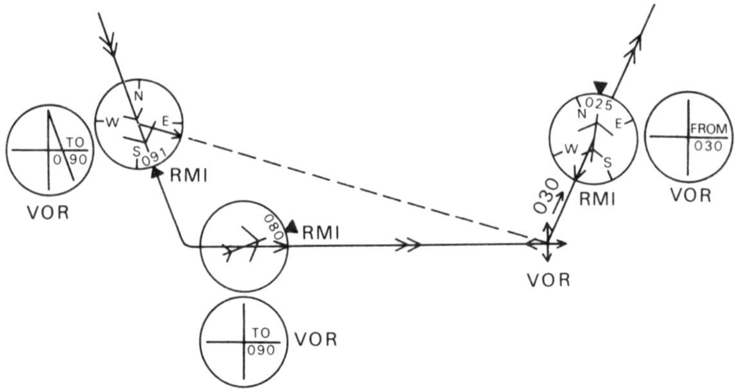

Fig. 7.7 Homing procedures using VOR L/R deviation indicator and RMI.

(b) If 055 was selected, the indication will be turn right (see Fig. 7.8).

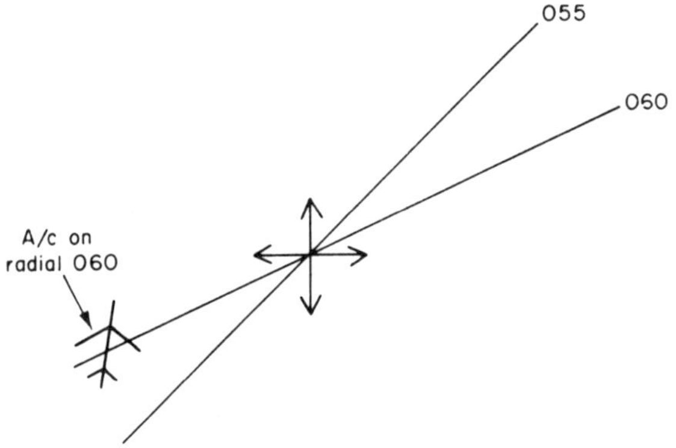

Fig. 7.8 Right turn indication.

(2) An aircraft is heading 060°(T), tracking on a VOR radial 258 Var 10°W. What should ADF read on beacon sited at VOR station?

Hdg(M) = 060°(T) + 10°W
= 070°(M)

Aircraft's track is 078°(M) ∴ aircraft has 8° starboard drift.

∴ NDB will read 360 + 8 = 008°

(3) An aircraft is heading 060°(T). Variation is 10°E. An NDB bears 200°(R). Show by means of a sketch how this information would appear on the face of RMI.

Rel Brg = 200
Hdg(M) = 050
QDM = 250 (See Fig. 7.9)

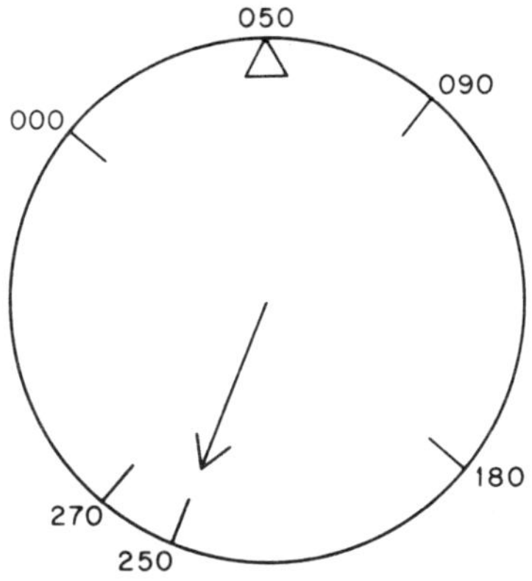

Fig. 7.9 Heading 050°(M) QDM 250.

(4) An aircraft is flying a constant heading with 8°P drift and is making good a track parallel to the centre line of an airway, but 5 nm off to the left of the centreline. Estimate the ADF reading of an NDB sited on the centreline of the airway, but 30 nm ahead.

In Fig. 7.10, the angle between the aircraft's track and the radio station is worked out using the 1 in 60 rule.

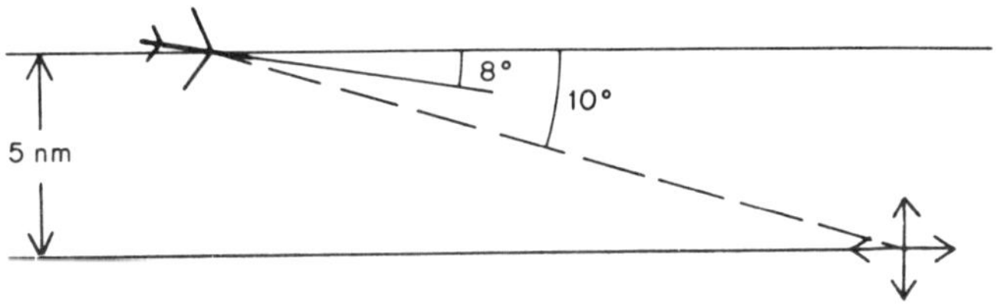

Fig. 7.10 Applying 1 in 60 rule.

$$\text{Tr Error} = \frac{\text{Dist off} \times 60}{\text{dist to go}} = \frac{5 \times 60}{30}$$
$$= 10°$$

The nose of the aircraft is offset to the right (to compensate for port drift) by 8°. Therefore, the angle between the nose of the aircraft and the NDB

$$= 10 - 8$$
$$= 2°$$

The ADF will read 002°.

(5) An aircraft is heading towards VOR, A, maintaining radial 130°, drift 10°P. NDB, B is due east of VOR, A and when it is abeam present track, heading will be altered for B. Drift will then be 5°S.

On an RMI show the approximate indications of A and B as they would appear

(a) shortly before altering heading for B and
(b) shortly after the heading has been altered.

See Fig. 7.11.

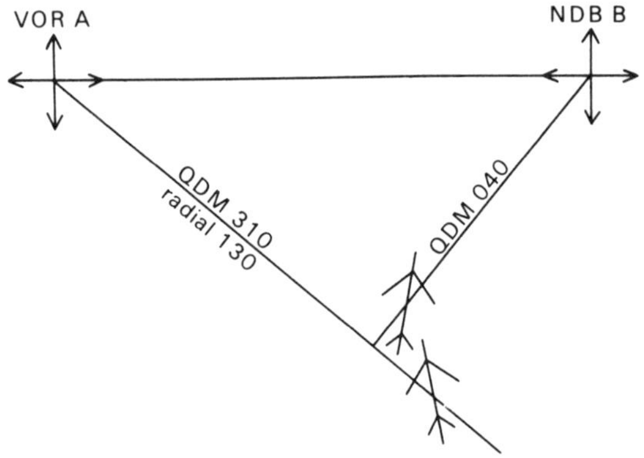

Fig. 7.11 On A's radial 130, abeam B.

Indications will be as follows:

(a) Just before altering heading:
RMI heading – 320(M)
VOR pointer – 310 (QDM to VOR)
NDB pointer – 035° (QDM from abeam position is 040°).

(b) After altering heading:
RMI heading – 035°(M)
NDB pointer – 040 (QDM to NDB, B)
VOR pointer – 300.

(6) Complete the following table:

	Hdg(M)	Rel Brg	RMI indication
(a)	210	182	–
(b)	130	–	160
(c)	040	–	160
(d)	–	240	038
(e)	–	193	026

Answers: (a) 032 (b) 030 (c) 120 (d) 158 (e) 193.

(7) An aircraft is homing to a VOR. Drift is 8°P, variation at VOR is 5°W, variation at the aircraft position is 4°W. Give the initial heading to maintain a radial of 253°

Answer: 081°(M)

(8) An aircraft's RMI is being fed with information from a VOR and an NDB. The aircraft is tracking to VOR on radial 228°. Drift is 5°P. An NDB bears 060°R from the aircraft. Give the two indications on an RMI.

Answer: VOR pointer 048°; NDB pointer 113°; RMI heading 053°(M).

(9) An aircraft is heading 041°(M). It is 20 nm from a VOR and on a bearing of 215°(M) from the station. The pilot has selected 212° on OBS and the TO/FROM indicator is indicating FROM. What information is displayed on (a) RMI (b) ILS/VOR indicator?

Answer: On the RMI, the heading is 041°(M) and the VOR pointer indicates its present QDM which is (215° − 180°) = 035°.

On the L/R indicator, the selected bearing is 3° off from his own radial, and to his right. Therefore the indications are: $1\frac{1}{2}$ dots in five dots or slightly over one dot in four dots. As the aircraft is heading against the selected radial and FROM is indicated, the left/right display will give him a left turn.

Chapter 8
INSTRUMENT AND MICROWAVE LANDING SYSTEMS (ILS AND MLS): FAN MARKERS

ILS is a pilot-interpreted runway approach aid, developed during the Second World War and now in world-wide use. The system provides the pilot with visual instructions enabling the aircraft to be flown along a predetermined flight path to the threshold of the runway being served by the system. In practice, the pilot descends to his decision/critical height and then by visual reference makes his final decision to land or overshoot. With this system the ground transmissions are continuous and no assistance from the ground control is required.

The runway being served by the ILS or precision approach radar (PAR) is called a precision instrument runway.

The ILS ground installation consists of the following three components:

(1) Localiser transmitter, together with its aerial system; this transmitter supplies approach guidance in azimuth along the extended runway centreline
(2) Glidepath transmitter, together with its aerial system; this transmitter provides approach guidance in the vertical plane
(3) Two or three marker beacons, each with its own aerial; they provide range check points

A layout of the ground system is shown in Fig. 8.1.

The localiser transmitter

The radio signals transmitted by the localiser antenna produce a composite field pattern along the approach direction, consisting of two overlapping lobes. The transmitter aerial is located in line with the runway centreline, approximately 300 m from the upwind end of the runway. The two lobes are transmitted on a single ILS frequency in VHF and in order that the receiver can distinguish between them, they are differently modulated. That lobe on the right-hand side of the runway as seen by the pilot making an approach is modulated by a 150 Hz signal and the sector it forms is called the blue sector (see Fig. 8.2). The lobe on the left-hand side is modulated by a 90 Hz note and the sector formed by it is called the yellow sector. An aircraft approaching the runway in the landing direction will detect more of the 90 Hz

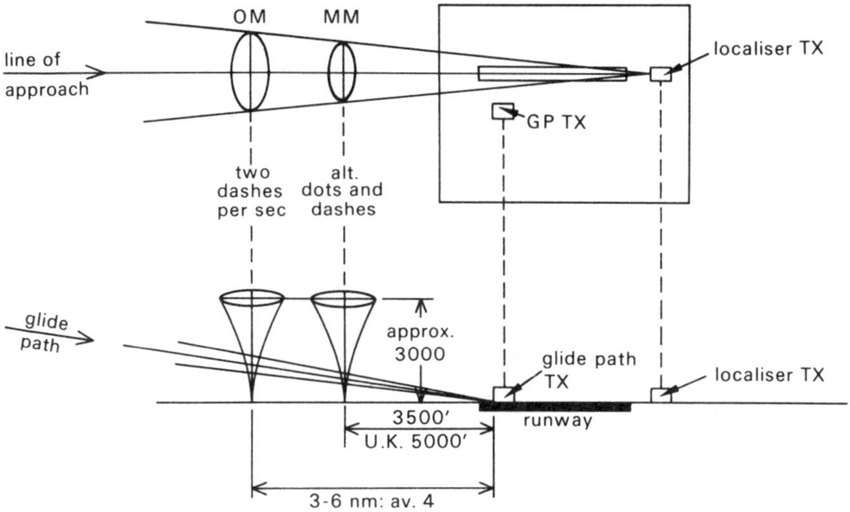

Fig. 8.1 ILS ground system layout.

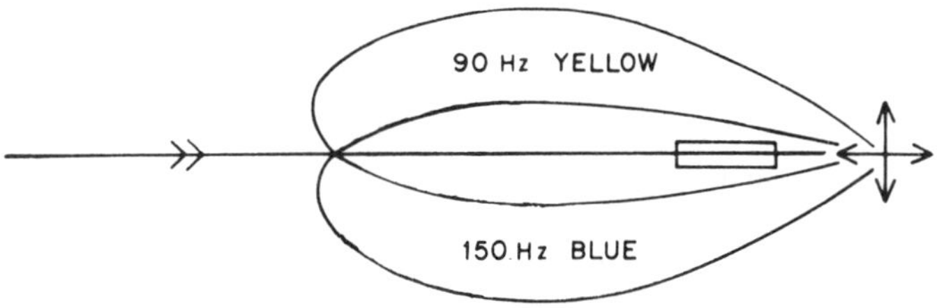

Fig. 8.2 Blue and yellow sectors.

modulation note and relatively less of the 150 Hz modulation note if it is to the *left* of the centre line. This excess of 90 Hz modulation (difference in depth of modulation (DDM)) will energise the vertical needle of the ILS meter (or VOR/ILS indicator) to indicate a right-hand turn. Similarly, an aircraft flying to the right-hand side of the centreline will have an excess of the 150 Hz modulation note and the needle will indicate a left-hand turn. The line along which the DDM is 0 defines the runway centreline. When flying along this line there will be no deflection of the needle, indicating that the aircraft is on the centreline.

Localiser coverage – category 1

The localiser coverage extends from the transmitter to 25 nm, 10° either side of the centreline. However, it widens to 35° from the centreline to a range of 17 nm (see Fig. 8.3). These dimensions may be reduced where it is necessary for topographical reasons. The vertical coverage in the areas already described is 7°. In this volume of airspace the radiation field strength is

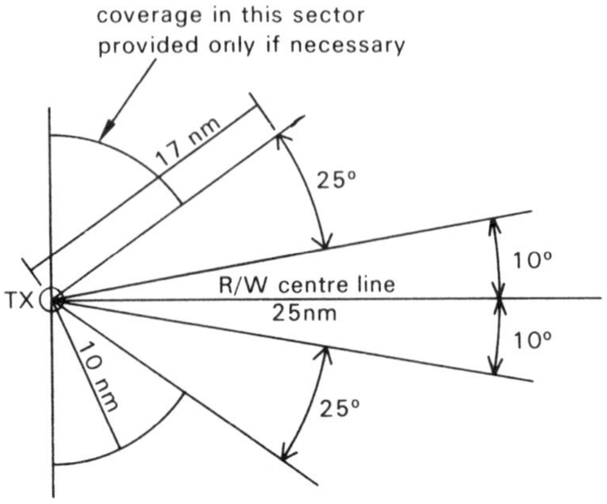

Fig. 8.3 Localiser coverage.

sufficient to permit satisfactory operational use of the localiser. (The minimum prescribed is 40 microvolts per metre.) Signal strength reduces rapidly outside this airspace where false localiser signals with reversed sense can occur: similarly, the maximum field strength is directed on the centreline, to a distance of 10 nm.

If the localiser centreline is being used for navigational purposes (e.g. taking a position line) it should be noted that the localiser signals are protected from interference out to a range of 25 nm at an altitude of 6250 ft along the on-course line. As for the accuracy, they are checked up to 10 nm.

The glidepath transmitter

Ideally this transmitter and its aerial should be located at the touchdown point on the runway. But in the early experimental days it was found that the personnel manning the transmitter did not run fast enough to get out of way and consequently kept damaging the landing aircraft. Hence it is located to one side of the runway, approximately 150 m from the centreline, 300 m upwind from the threshold. The transmission is beamed in the vertical plane in two lobes similar to the localiser transmission. The upper lobe has a 90 Hz modulation, the lower lobe has a 150 Hz modulation. The line along which the two modulations are equal in depth defines the centre line of the glide slope. It is generally 3° from the horizontal but it could be adjusted to between 2° and 4° to suit the particular local conditions (see Fig. 8.4).

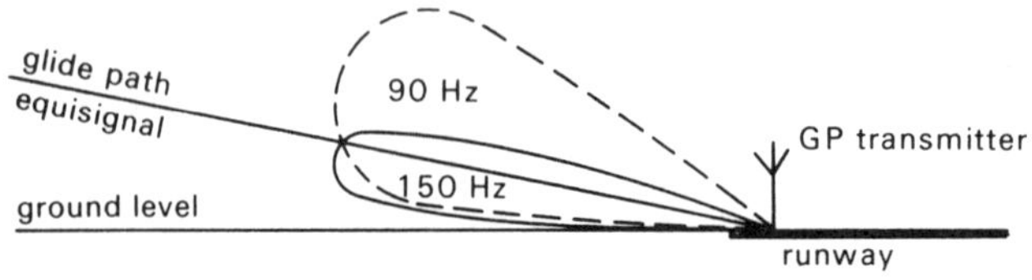

Fig. 8.4 Glide slope pattern.

The glidepath (GP) coverage – category 1

The coverage in azimuth extends 8° on either side of the GP centreline, to a distance of 10 nm (see Fig. 8.5). In the vertical plane the coverage begins from 0.45 × GP angle (θ) above the surface to 1.75 × GP angle (θ) above the surface. This means that for a 3° GP angle the coverage is from 1.35° to 5.25° above the surface. (Note: The coverage and field strength data given above for localiser and glide path transmissions are appropriate for category 1, ILS.)

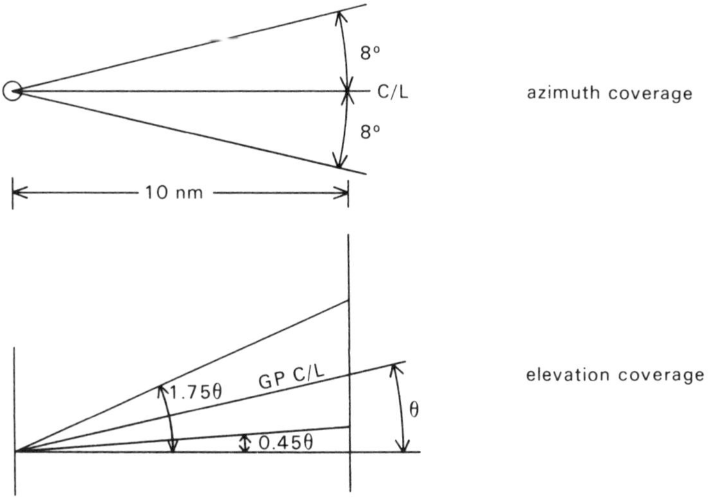

Fig. 8.5 Glidepath (GP) coverage.

ILS indicator

ILS uses the VOR's L/R deviation indicator (OBS being inoperative), incorporating an additional horizontal needle. This needle is inoperative when the indicator is displaying VOR information. The indicator is also described as the VOR/ILS meter (Fig. 8.6). The indicator illustrated in Fig. 8.6 is a five-dot indicator, dot 1 being the outer edge of the centre circle.

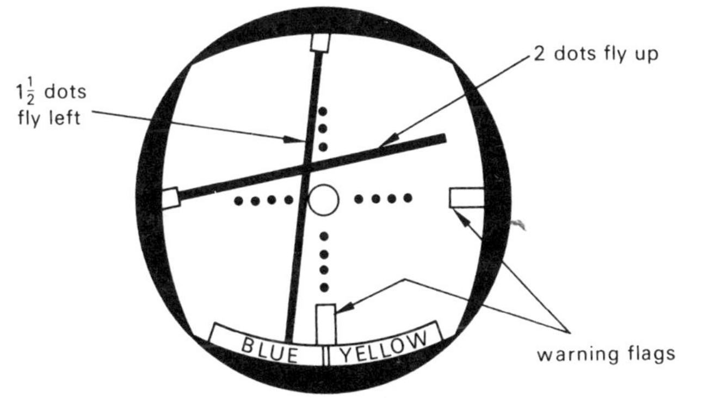

Fig. 8.6 ILS indicator.

When used with ILS, the vertical pointer indicates the aircraft's deviations in azimuth, and the horizontal needle indicates its position with regard to the glide slope centreline. Both needles remain in the central position when

- the receiver is switched off or no signals are being received, or
- the aircraft is on the centre line of the localiser and the glide path.

The bottom of the dial is coloured blue to the left and yellow to the right. This indicates to the pilot the localiser sector he is in, e.g. if the vertical needle has swung to the left, the aircraft is in the blue sector.

All ILS indicators employ two failure warning flags; one operating in association with localiser signals (this flag also operates in association with VOR), the other with the glidepath signals. They fall into view in the windows and the needles return to the central position when:

- ground or airborne equipment has failed or is switched off, and
- out of the service area or the signals being received are too weak. (The signal strength falls off quickly once outside the service area.)

Monitoring of ILS transmissions

Both localiser and glidepath transmitters are automatically monitored by monitoring equipment located in an area of guaranteed reception within the normal service sector. It will act in one of the following circumstances:

(1) a localiser shift of more than 35 ft from the centreline
(2) a glide slope angle change of more than 0.075 × basic glidepath angle,

e.g. $$3 \times 0.075^\circ = 0.225^\circ$$

(3) a reduction in power output of 50% or more of any of the transmitters.

In any of the above circumstances, the monitoring unit will provide warning to a designated control point and cause any of the following to occur before a standby transmitter is brought into use:

(1) the cessation of all radiations
(2) the removal of the ident signal and/or the navigational component (i.e. localiser and glide path)
(3) if the ILS is category 2 or 3, the monitor may permit operation to a lower category, i.e. 1 or 2. (See later reference to ILS categories.)

Localiser indications

The vertical pointer is used for localiser indications. The needle tells you which way to turn and the horizontal deflection scale gives you an estimate of the angular displacement from the centreline. The coloured sectors at the bottom tell you which sector you are in. Follow the illustration in Fig. 8.7.

Aircraft A is in the blue sector, and the needle indicates left turn. Aircraft B is in the blue sector, and the needle indicates left turn. Thus, the indication is given according to the sector the aircraft is in, not according to its heading.

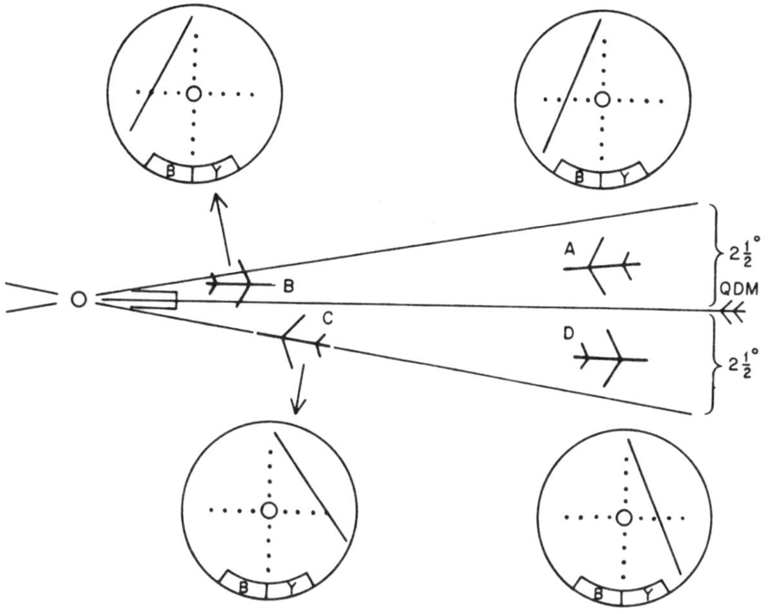

Fig. 8.7 Localiser indications.

In this case aircraft B is on the right-hand side of the centre line and on reciprocal heading, therefore it will have to *reverse* the indication. The same applies to aircraft D which is on the left-hand side (or yellow sector) with reciprocal heading. Its turn right indication is reversed if it wishes to regain the centreline. In all cases notice that the needle indicates the sector the aircraft is in and 'follow the needle' rule applies when making an approach. Reverse the indication if going away on the QDR.

As for the deviation scale, presentation of the centreline beam is 5° wide, that is, $2\frac{1}{2}°$ on either side of the centreline. Maximum deflection of the needle occurs when the aircraft is $2\frac{1}{2}°$ or more from the centreline. On a four-dot indicator, one dot represents a deviation of approximately 0.6°, on a five-dot indicator, 0.5°. (Remember that the same needle will give a full deflection when a 10° deviation from a VOR radial occurs.)

Glidepath indications

The horizontal needle is used in conjunction with the glidepath transmissions. If the aircraft is below the glidepath the needle moves upwards, indicating that the aircraft should fly up to regain the glideslope. This indication will occur irrespective of the heading, that is, whether the aircraft is on QDM or QDR. Therefore, a departing aircraft wishing to climb along the glideslope will obey the needle. If an aircraft (approaching or departing) is above the glidepath the needle will move downward, indicating that the pilot should come down (see Fig. 8.8).

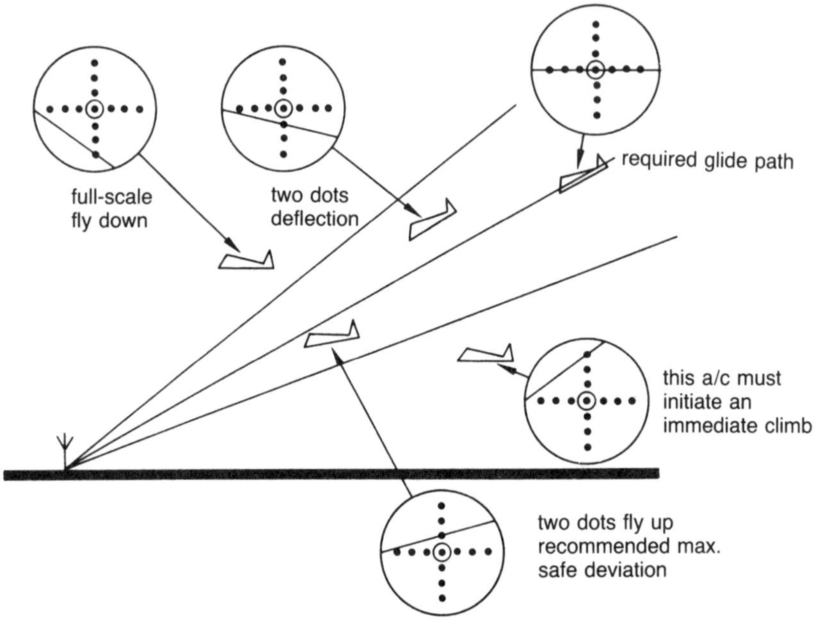

Fig. 8.8 Glidepath indications.

Full deflection of the needle occurs when the aircraft is 0.7° or more above or below the glidepath (1 dot = 0.14°). A two-dot fly-up indication out of four dots or $2\frac{1}{2}$ dots out of five dots (in other words, half full deflection) is to be regarded as the maximum safe deviation below the glidepath. On seeing any indications below this, an immediate climb must be instituted: remember this at all costs.

In Fig. 8.6, the indicator gives the combined indications of localiser and glidepath deviations. The interpretation depends on whether you are approaching the runway (on QDM) or going away from the runway (QDR):

- if on QDM, the indications in Fig. 8.6 instruct you to turn left and climb
- if on QDR, the instructions are to turn right and climb.

Marker beacons

Usually two, some times three marker beacons are installed along the extended centreline to give range indications on approach. This enables the pilot to check his height as he passes each marker. All markers transmit on a single frequency of 75 MHz and radiate a fan pattern upward to a calibrated height of approximately 3000 ft. The marker farthest from the touchdown point is placed approximately 3 to 6 nm, average 4 nm, from the touchdown point and is known as the outer marker (OM). It transmits a low-pitched 400 Hz modulation signal and identifies itself in morse as well as visually. When crossing the beacon, a series of dashes is heard in the earphone, the rate being *two dashes* per second. Simultaneously the *blue* marker light will flash dashes at the same rate.

The next marker on the approach path is called the middle marker (MM), placed approximately 3500 ft (5000 ± 500 ft in the UK) from the touchdown point. It transmits a series of alternate dots and dashes at a higher pitch, 1300 Hz, which are heard in the earphones and also seen flashing on the *amber* marker light.

The marker nearest to the beginning of the runway is called the inner marker (IM). It transmits six high-pitched (3000 Hz) dots per second and the *white* light flashes. *When installed*, it is located between 250 and 1500 ft.

Summary of the markers associated with ILS

Designation	Distance from the RW touchdown	Signal characteristics
Outer marker	3–6 nm, average 4 nm	Transmission modulated by 400 Hz, 2 low-pitched dashes per second – blue light flashes
Middle marker	3500 ft (5000 ± 500 ft in the UK)	1300 Hz signal keyed to form alternate dots and dashes – amber light flashes
Inner marker	250–1500 ft	3000 Hz signal, 6 high-pitched dots per second – white light flashes

Thus the markers are identified in three ways: by audio signals, visual signals and the transmission pitch.

One or two locators may be used to supplement the ILS. These locators are low-powered NDBs and share the sites of the OMs and MMs. If only one locator is used, it is usually installed on the site of the OM. The transmission frequencies of these locators (where two are being used) should not be closer than 15 kHz, otherwise mutual interference may result. Also, they should not be further apart than 25 kHz to permit a quick tuning shift when operating on a single indicator. These locators serve a three-fold purpose:

- they assist the pilot to home to the station and subsequently join the ILS pattern
- they may be used for holding purposes
- they provide a double check when passing over the markers.

On approach charts they are indicated by a standard abbreviation LOM (locator, outer marker) and LMM (locator, middle marker).

DME is used as an alternative to the markers. Its radiation is then so adjusted as to give zero range at or near the touchdown point. It may be frequency-paired with the ILS localiser so that when the ILS is switched on, the DME automatically starts functioning. One DME may serve both approaches to a runway.

Airborne equipment

The airborne equipment consists of:

- channel control box
- VHF localiser receiver
- UHF glidepath receiver
- 75 MHz marker beacon receiver
- ILS meter or VOR/ILS indicator
- three separate aerials for the three receivers.

A block schematic diagram of the airborne equipment is shown in Fig. 8.9.

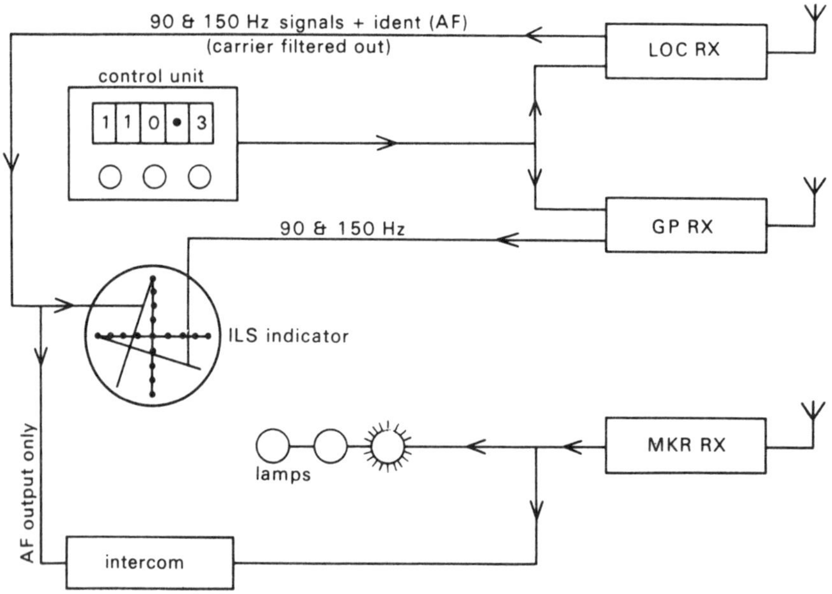

Fig. 8.9 Airborne equipment.

Frequencies

Localiser. Frequencies allocated to ILS in the VHF band are: 108 to 112 MHz at odd 'first' decimals. e.g. 108.10, 108.15. In the UK the military uses some even decimals as well. Future ILS frequency assignments will be on frequencies ending in odd tenths plus a twentieth of a MHz.

Glidepath. Transmission takes place in the UHF band on 20 spot frequencies from 329.3 MHz to 335 MHz at 300 kHz spacing, e.g. 329.3, 329.6, 329.9, etc. The use of UHF is to produce more accurate beams. In the future, glidepath channel spacing will be reduced from 300 kHz to 150 kHz.

Frequency pairing. Localiser and glidepath transmissions are frequency-paired. This means that for each one of the twenty localiser spot frequencies there is one glidepath frequency allocated to it. For example, frequency of 109.3 is paired with 332.0 and 111.5 is paired with 332.9. (You need not memorise these paired figures – they are for illustration only.)

The advantages of frequency pairing are as follows:

(1) By means of one switch, two receivers are activated – this reduces the workload.
(2) Frequency selection is quicker and easier. There is no need to look up the glidepath frequency in the flight information documents.
(3) A potential error in frequency selection is prevented.
(4) Separate identifications are not necessary.

Type of emission

The type of emission is A8W for localiser, glidepath and marker transmissions.

Identification

As localiser and glidepath frequencies are paired, whenever the localiser frequency is selected, the glidepath receiver circuits corresponding to the paired frequency are automatically energised. Therefore, if you are subsequently receiving glidepath signals, they can only be from the correct transmitter. Hence it is unnecessary for both localiser and glidepath transmitters to identify themselves separately.

The ident takes place on the localiser transmission. Its carrier is amplitude-modulated by a horizontally-polarised 1020 Hz tone to give the ident. The ident itself is by two, three or more letters in morse, seven words per minute. Where it is necessary to distinguish an ILS quickly from other facilities, the ident may be preceded by the letter I. And since the localiser carries the ident, if it becomes unserviceable or it is withdrawn from service for any reason, the ident will be automatically suppressed.

Ground-to-air voice communication may be conducted on category 1 and category 2 ILS localiser carriers provided it does not interfere in any way with the navigational or ident function of the localiser.

ILS reference datum

This is defined as a point at a specified height (usually around 50 ft) located vertically above the intersection of the runway centreline and the ILS landing threshold through which the downward extended path portion of the ILS glidepath extends.

ILS categories

Background. The ILS project, originally conceived to develop a blind landing system, did not quite reach its objective and turned out to be an instrument 'approach to landing' aid. But still it was a great step forward in those days: its faults were forgiven by the operators and it received ICAO's blessing in 1946.

As civil aviation developed, the operators became increasingly more weather-conscious. They disliked the thought of delaying a flight or wasting time and fuel while holding overhead an aerodrome waiting for the weather to clear. The ILS had its faults, the main one being production of bends in the beam. These were produced by reflections from obstacles on and around the aerodrome, e.g. airport structures, vehicles, aircraft flying overhead the localiser aerial, and so forth. The airborne equipment, similarly, was just adequate to handle the existing system.

In 1958 British Airways (then BOAC) announced its intention to go for all-weather-operation and a positive move in that direction began. Improvement had to come to both the ground and the airborne equipment. As for the ground equipment, it was decided to develop an entirely new landing system based on modern technology, but in the meantime to retain and improve the system. For an improvement, new transmission data were prescribed, course structures and course bends were tightly defined, the forward beam was narrowed down to reduce the reflections and, to assist in the overall advancement, the airport and environment needed to be 'cleaned up' from interference. As the improvement progressed, a system of categories was established to define the capability of a particular ILS. As a matter of interest in the UK, ILS serving runway 10L (now 09L) at London Heathrow was the first one to be upgraded to category 2. These categories are called ILS facility performance categories, and they are defined as follows.

ILS facility performance categories

Category 1 – an ILS capable of providing accurate guidance from the coverage limit down to a height of 200 ft above the ILS reference point

Category 2 – an ILS capable of providing accurate guidance from the coverage limit down to a height of 50 ft above the ILS reference point

Category 3 – an ILS capable of providing accurate guidance from the coverage limit down to the surface of the runway.

Operational performance categories

We saw from the above that with an improved ground equipment, guidance down to the surface became possible. The operational objective of establishing the above categories is defined by ICAO in terms of operational performance categories (also known as operational approach categories or weather categories). The criterion here is the corresponding improvement in the airborne equipment.

Although the transmitted signals might be absolutely correct, a receiver with out-of-balance components can produce an indication of false centreline, both in elevation and in the azimuth. Consequently, a pilot descending with his ILS needles perfectly centralised, on being visual, may find himself displaced to one side or the other or too high/too low. The improvement had to start from here; the search for the components that would reproduce information faithfully and reliably was on. In fact, the improvement in airborne equipment went ahead side by side with ground equipment and full 'hands off' landing tests were being carried out as far back as 1961. Now with super, complex, computer-controlled equipment on board, an aircraft may be certificated to an appropriate category from the following classification.

Category 1 – A precision instrument approach and landing with a DH not lower than 60 m (200 ft) with RVR 550 m

Category 2 – A precision instrument approach and landing with a DH lower than 60 m (200 ft) but not lower than 30 m (100 ft) with RVR 300 m

Category 3 – A precision instrument approach and landing with a DH, if any, lower than 30 m (100 ft) and an appropriate RVR.

where RVR is Runway Visual Range and a Precision Approach is an Instrument Approach to landing using ILS, microwave landing system (MLS) or PAR for guidance in both azimuth and elevation.

Back beam

The localiser transmission is normally directed in the direction of the approach area to provide azimuth guidance to the approaching aircraft. But usually there is a certain amount of overspill of radiation behind the localiser aerial and the signals would be received when flying in this area. This beam is not to be used. Some transmitters are, however, designed to radiate a back beam. Where this facility exists, it can be used when overshooting the precision runway. It can also provide a back course approach to the reciprocal runway. It must be noted that when using a back course there is no benefit of a glidepath. Usually, they are less accurate than the front beams, there are no range-check markers and they are not checked for accuracy. The needle sense is reversed.

ILS offset localiser signals

Occasionally for technical reasons a localiser aerial has to be temporarily offset to one side of the centreline. On these occasions the relevant information is published in NOTAMS. If the need for an offset aerial is to extend over a period, the information is published in the RAC section of the UKAIP.

Under the early system, with an offset localiser the DH occurred at that height at which an aircraft on the glide path would transit the middle marker. Under the present system, when the offset does not exceed 2°, the DH is calculated from the published obstacle clearance limit (OCL) or 200 ft above the runway threshold, whichever is higher.

False glidepaths

These are defined as those loci of points in the vertical plane containing the runway centreline at which the DDM is zero, other than that locus of points forming the ILS glidepath.

What all this means is that, in the process of producing the glidepath, due to the inherent metallic structures at the point of transmission, and the aerial's propagation characteristics, the radiated twin lobes are repeated several times above the true centreline. These produce several other equisignals (see Fig. 8.10). The number of such false glidepaths produced at any ILS site depends on several factors such as the design of the transmitting aerials, obstructions around the transmitter, transmission power and such like. These false glidepaths, however, are not a danger to the pilot for the following reasons:

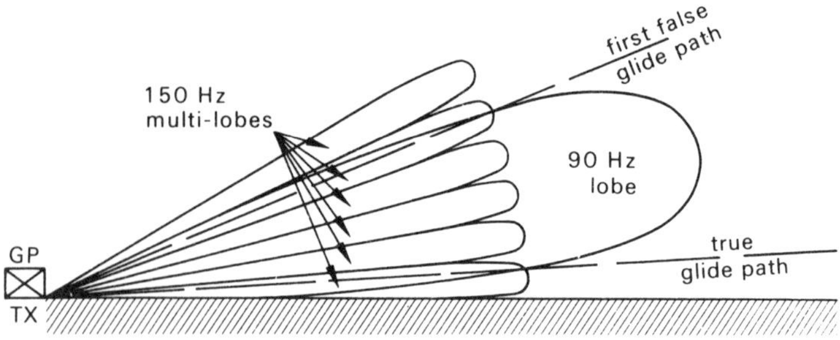

Fig. 8.10 False glidepath.

(1) The first false glidepath does not occur until above 6°. Thus, if you caught it, you would soon appreciate the mistake.
(2) False glidepaths always occur above the true glidepath, and therefore cannot bring the aircraft dangerously low.
(3) It is a normal practice, when intending to carry out an ILS approach, to establish on the localiser first and then to meet the glideslope from underneath. It is most unlikely that a pilot would miss the true glideslope and continue flying level until the next equisignal is reached.
(4) With the recommended localiser coverage in elevation of 7° and the glidepath coverage of 1.75 × GP angle, the signals being received on the false glidepath will be weak and the warning flags may operate.
(5) It is a recommended practice that establishment on the promulgated glidepath be confirmed by the relationship between aircraft height and the distance to the runway threshold.

Pilot's serviceability checks

During an approach the localiser and glidepath serviceability may be checked by the pilot in two ways:

(1) The failure warning flags should remain clear of the window. The warning flags are actuated by the sum of the two modulation depths, and as we saw earlier, in the case of total unserviceability, the monitor removes a navigational component. As soon as this happens, the flag will appear in the window.
(2) A pilot monitoring the identification signals will soon be warned if the ident signals stop coming.

As a further precaution, where a ground precision approach radar is available, it is mandatory for the radar to monitor ILS approaches in certain weather conditions.

ROD and other calculations

While carrying out an ILS let-down, it will be necessary to calculate the ROD for the glidepath angle. This is calculated in ft/min from the following formula:

$$\text{ROD} = \text{GP angle} \times \frac{\text{GS(kt)}}{60} \times 100$$

Example: calculate the rate of descent (in ft/min) for a glidepath angle of 2.9° and a ground speed of 112 kt

$$\begin{aligned}\text{ROD} &= 2.9 \times \frac{112}{60} \times 100\\ &= 541\,\text{ft/min}\end{aligned}$$

With ILS, the aircraft height, the ground distance to go and the glideslope angle make a right-angled triangle. Consequently, if two of the above factors are known, the third one can be calculated. In the absence of maths tables, and in any case for the practical usage, the 1 in 60 rule may be used to solve the problems.

Examples

(1) What is the approximate height of an aircraft at 2 nm range on a glideslope of 2.7°?

$$\text{TE (track error is our GP angle)} = \frac{60 \times \text{ht(ft)}}{\text{dist to go (ft)}}$$

$$2.7 = \frac{60 \times \text{ht(ft)}}{2 \times 6080}$$

$$\therefore \quad \therefore \text{Ht} = \frac{2.7 \times 2 \times 6080}{60}\,\text{ft}$$

$$= 547\,\text{ft}$$

(2) At a distance of 3 nm from the threshold a pilot receives a full deflection on the glidepath pointer, indicating fly up. Approximately how many feet below the glidepath is the aircraft at this time?

The maximum deflection occurs when the pilot is 0.7° below the glidepath. Therefore,

$$\text{vertical distance from the centreline} = \frac{0.7 \times 6080 \times 3}{60}\ \text{ft}$$
$$= 213\ \text{ft}$$

(3) An aircraft on an ILS approach indicates half full-scale deflection on the glideslope pointer, giving fly-up indication. At 2.2 nm range from the threshold there is an obstruction, 285 ft above threshold level. What will be the vertical clearance from the obstacle when the aircraft passes over it on a 3° glideslope?

Half full-scale deflection occurs when the aircraft is 0.35° below the glidepath. Thus, the aircraft is (3° − 0.35°) = 2.65° above the surface at 2.2 nm range. At this point

$$\text{height of aircraft} = \frac{2.65 \times 2.2 \times 6080}{60}$$
$$= 590.8\ \text{ft}$$

$$\text{The vertical clearance of the aircraft from the obstacle} = (590.8 - 285)\ \text{ft}$$
$$= 305.8\ \text{ft}$$

Limitations of ILS

ILS has the following limitations:

(1) Signal corruption causes unpredictable bends in localiser and glideslope beams. The pilot must always remain on the alert and particularly so when making a fully automatic landing.
(2) Ground effects restrict its use particularly in mountainous and other difficult areas.
(3) It entails too large permanent fixtures for each runway.
(4) For reasons of minimising interference, the landing rate is kept low, and there may also be restrictions of vehicle movement on the ground.
(5) Only a limited number of channels (40) are available and the effectiveness of even these may be reduced in future by interference from FM broadcasts.

The CAA normally has a pink AIC current on the use of ILS in the UK.

Microwave landing systems

Because of the limitations above, and also because helicopters and short take-off and landing (STOL) aircraft using ILS have to conform to the patterns flown by large fixed-wing aircraft, a more readily usable system has been developed.

Originally a world-wide standard MLS was planned to be progressively implemented so as to become the primary approach and landing aid from the year 2000 onwards. However, the subsequent development of DGPS has led certain countries (in particular the USA and Canada) to abandon MLS in favour of DGPS as the landing aid of the future (see Chapter 20).

In contrast to the ILS principle, which embodies a localiser and a glideslope providing a clearly defined approach path above the runway's extended centreline, MLS allows approaches anywhere within its horizontal and vertical fan-shaped coverage area. See Fig. 8.11.

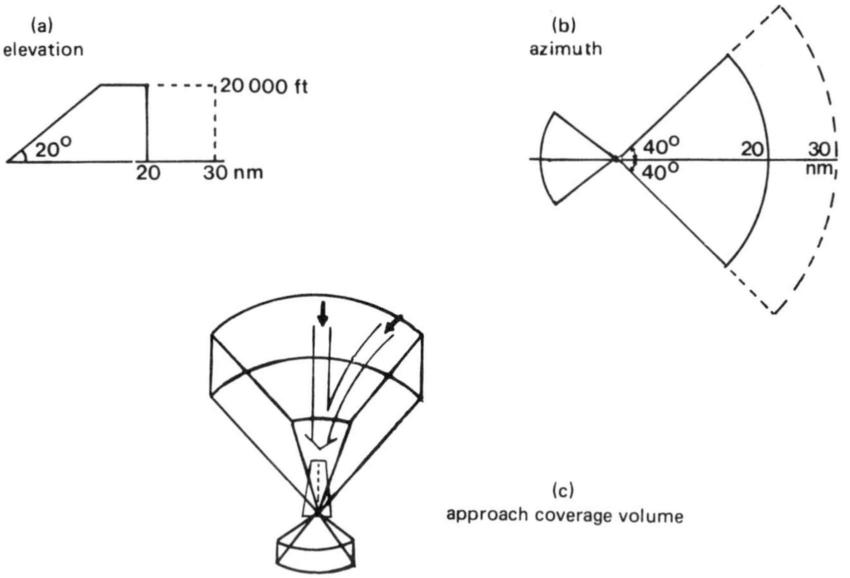

Fig. 8.11 Microwave landing system (MLS).

The system has an azimuth transmitter (corresponding to ILS's localiser) which provides a fan-shaped horizontal approach zone, usually ±40° of the runway centre-line. Similarly the MLS elevation transmitter (corresponding to the glide slope on ILS) produces a fan-shaped vertical approach zone usually ranging from 0.9° to 20°. It is possible to provide even steeper approach angles but because of aircraft handling problems it is not envisaged that even the upper part of the 20° sector will be used.

There will be a DME facility (corresponding to the marker beacons on an ILS approach) and sometimes a back azimuth.

With MLS, a time reference scan beam system (TRSB) is used to determine the aircraft's position. Its transmitters produce narrow-width beams which sweep to and fro through the 80° azimuth and 19° vertical ranges. The aircraft receiver measures the time interval between sweeps to determine position and the pilot can select an appropriate approach path. The receiver then creates an ILS-like localiser for the chosen approach.

By this means a helicopter can make an approach from, say, 35° to the runway on a 6° elevation immediately after a conventional big jet has approached along a standard 0° azimuth 3° elevation approach.

MLS also requires less space for the ground equipment, has 200 channels available and does not suffer from ground effects, as it operates in the gigahertz (or radar) frequency.

Advantages

Compared to ILS, MLS have the advantages of:

- Greater accuracy and reliability due to less signal corruption from interference
- Extremely good guidance capacity
- Insensitivity to geographical site, which enables it to be established where an ILS installation cannot be accommodated
- More channels available (200)
- Very wide three-dimensional coverage, allowing curved flight path captures and final approaches on different glide slopes
- Better means of controlling and expediting aircraft movements in terminal areas.

Fan markers

Fan markers transmit a narrow vertical fan-shaped beam of horizontally polarised radiation. All markers operate on a single frequency, 75 MHz. Because of the shape of the transmission, they cannot be heard unless the aircraft is in the fan, and therefore, they cannot be used as directional aids.

Fan markers have two main uses: they are used to mark reporting points and they are also used in conjunction with ILS to provide a precision approach facility.

At a reporting point, a fan marker is identified by a high-pitched (3000 Hz) audio signal giving out identification in morse, 6 to 10 words per minute A2A emission. Further, the white light in the airborne installation flashes to identify visually. The vertical coverage of the fan is limited to the operational requirements: there are low power fan marker beacons and high power beacons. Because in the horizontal plane the area of coverage increases with height, if accurate navigation is required, the time of entering and leaving the fan should be noted and the mean time taken for the fix. On some equipment a high/low switch is fitted which may be used to reduce the coverage area inside the fan.

Chapter 9
BASIC RADAR

When the word 'radar' was coined, it expressed precisely the function it performed, that is, radio detection and ranging. The use of the pulse technique clearly distinguished it from 'radio' which used continuous waves. With the subsequent advance in the technology, what we would have originally described as radio (a continuous wave) can now perform the tasks of radar – detecting and ranging (for example, a radio altimeter). Radar, in its turn, now performs a variety of tasks not included in its original definition. These tasks include turbulence indication by weather radar, navigational assistance from hyperbolic systems such as Loran, and ground speed and drift from Doppler. Radar may now perhaps be described as radio systems performing particular functions inside the range of the radio spectrum.

Radar frequencies

Radar occupies frequencies from VHF upwards. The reasons for the choice of higher frequencies are as follows:

(1) It gives freedom from external noise and ionospheric scatter
(2) Radar using the beam technique operates efficiently with narrow beams; these can be produced at shorter wavelengths
(3) Similarly, shorter pulses can be produced with shorter wavelengths
(4) The efficiency of reflection from an object depends on the size of the object in relation to the wavelength. At shorter wavelengths the signals will be reflected more efficiently by the reflecting objects.

Timing in radar

It will be appreciated that where timing is required to be carried out for radar operation, these times must be essentially very small, considering that a radio wave travels 300 000 km/sec. Fortunately, radar can measure these small time intervals very accurately. The times are measured in microseconds, occasionally in milliseconds.

$$1 \text{ second} = 1000 \text{ milliseconds or } 1\,000\,000 \text{ microseconds } (\mu s)$$

$$1\,\mu s = \frac{1}{1\,000\,000} \text{ second} = \frac{1}{1000} \text{ millisecond}$$

Radar systems work by a variety of techniques. Two such techniques that interest us are the pulse technique and the continuous wave technique. Of these two, the pulse technique is far more widely used and we will discuss this first.

Pulse technique

Primary radar, secondary radar and Doppler radar all employ the pulse technique although performing vastly different tasks. The technique involves transmission of energy not in the form of a continuous wave but in very short bursts. Each tiny burst of this CW is given a predetermined shape and radiates in the form of a pulse. The mechanism of pulse transmission is shown in Fig. 9.1. The duration or size of the pulse is called the pulse width or pulse length. Although the pulse width is very small it can contain many radio frequency cycles. For example, suppose a radar pulse is transmitted on a carrier frequency of 1000 MHz (DME). If the pulse length is 3.5 μs, the number of cycles of the carrier frequency that occur in each transmitted pulse may be calculated as follows:

number of cycles occurring in 1 second = 1 000 000 000
number of cycles occurring in 1 μs = 1000
and number of cycles occurring in 3.5 μs = 1000 × 3.5 = 3500.

Thus, each burst contains 3500 complete cycles and if the transmission was at 10 000 MHz, each pulse would contain 35 000 cycles. These figures give us some idea of the dimensions and magnitudes we are talking about.

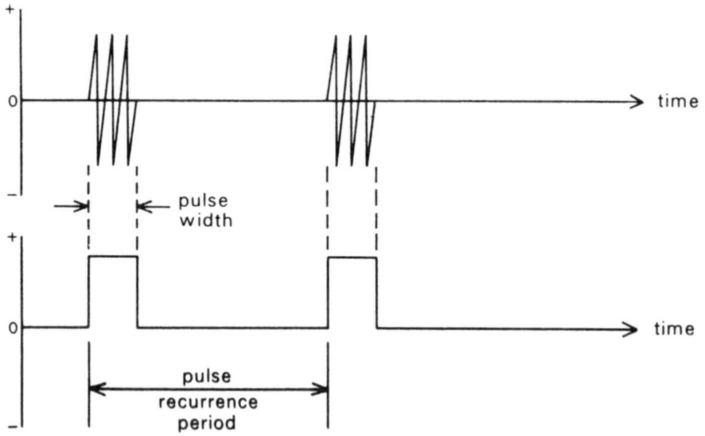

Fig. 9.1 Pulse transmission.

The distance between two pulses *in time* is called pulse recurrence period (PRP) or interval (PRI) and the number of pulses transmitted in one second is called pulse recurrence frequency (PRF) or rate (PRR). The relationship between these two terms is

$$\text{PRP} = \frac{1}{\text{PRF}} \text{ seconds}$$

Example: If the PRF is 250, what is the PRP of the transmission?

$$\text{PRP} = \frac{1}{250} = \frac{1\,000\,000}{250}\,\mu\text{s}$$
$$= 4000\,\mu\text{s}$$

Pulse shape. A pulse is given its shape by the process of pulse modulation and it is a design consideration. Although rectangular pulses can be produced by applying an instantaneous rise in the voltage, followed by an instantaneous collapse to zero, a practical pulse has a finite buildup time and decay time. The amplitude, pulse width, rise and decay times (all these factors defining a pulse) are subject to ICAO approval in respect of individual systems. A typical ICAO approved pulse as used in DME is shown in Fig. 9.2.

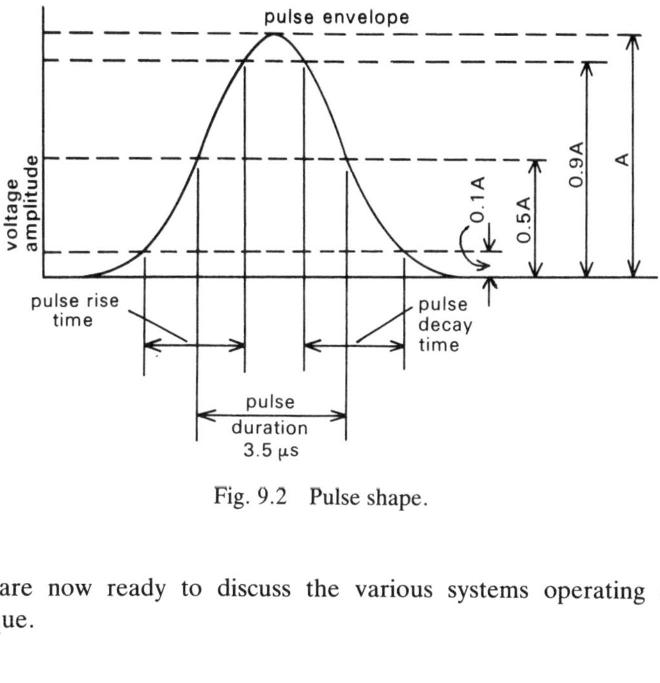

Fig. 9.2 Pulse shape.

We are now ready to discuss the various systems operating on pulse technique.

Primary radar

This is the original radar and uses the principle of pulse technique to determine range and bearing of an object. Working on echo and search light principle, a transmitter transmits a train of beamed pulses either in a fixed direction or omnidirectionally by a rotating scan in azimuth (surveillance radar). The beam may also scan in elevation according to the purpose of the equipment. All objects in the path of the pulses which are of a size commensurate with the wavelength, will reflect and scatter the energy. Some of this reflected energy will reach the receiver, but it will be greatly weakened. The strength of these echoes depends on several factors:

- power of the transmitter
- range of the reflecting object
- shape, material and attitude of the reflecting object
- size of the object in relation to the wavelength.

This reflected energy will be processed through the receiver and fed to the indicator in an appropriate form to give the information. In this process the object's co-operation is not required.

Distance measurement – echo principle. Radar finds the distance of an object by timing the interval between the pulse's despatch and its return as an echo. This timing is done electronically (we will see later in this chapter how a cathode ray tube (CRT) could be utilised to do this) and knowing the speed of the electromagnetic waves, the formula, distance = speed × time, can be solved. It will be noticed that the distance found in this formula is the one-way distance to an object. As the pulse has travelled out and back, the range of the object is half the distance so found, or, range = (speed × time)/2.

Example: An echo registers a time of 500 microseconds. What is the distance in kilometres of the object reflecting it?

$$\text{Range} = 300\,000\,000 \times \frac{500}{1\,000\,000 \times 2}\,\text{m}$$

$$= 75\,000\,\text{m}$$
$$= 75\,\text{km}$$

An alternative way of expressing the formula for the speed of radio waves in terms of a nautical mile is:

1 nm in 6 microseconds or
$\frac{1}{6}$ nm per μsec.

In implementing the above principle, two assumptions are made: that the speed of the electromagnetic waves is constant and that the waves travel in straight lines. Neither assumption is valid in the earth environment but the variations are so small that they can be ignored without incurring sizeable penalty.

Determination of direction. The search light principle utilises transmission of radio pulses concentrated into a very narrow beam. The beam width should be kept as narrow as possible for accurate bearing discrimination. Narrow beams can be produced either by shortening the wavelength or by increasing the aerial size. With advanced techniques an aerial can be adjusted electronically to give a beam of the required width. The beam is made to scan through azimuth or elevation, starting from a fixed datum point. The direction of the object then is the direction in which the beam is pointing at the time when the echo is received. It is read from a scale, calibrated from the starting point.

Ranges of primary radar. The ranges available depend on numerous factors; of particular interest to us are the following:

(1) *Transmission power.* It is obvious that an increase in power will increase the range subject to altitude. However, with radar, the signal not only has to travel to its destination (reflecting object) but must travel an equal distance back to the receiver with sufficient strength to predominate the internal receiver noise. Thus, the power/range relationship for primary radar is given in the expression, max. range = $\sqrt[4]{(\text{power})}$. This means that the power must be increased 16 times to double the range.

(2) *Characteristics of the reflecting object.* The size and shape of the object, the reflecting material (metal will reflect more efficiently than wood), aspect of the target: these factors determine the strength of the echoes coming back. An aircraft reflecting from the length of the fuselage will give a better echo than when its nose and tail are in line with the incoming wave. Further, an aircraft in an unusual attitude may shift polarisation of the waves, causing polarisation fading at the receiver. 'Stealth' bombers are designed specifically to be poor reflectors.

(3) *Pulse recurrence frequency (PRF)/Pulse recurrence rate (PRR).* This determines the *maximum unambiguous range* of the equipment. Each pulse must be given time to travel out to the most distant reflecting object as planned, and return, before the next pulse goes out. Otherwise it will not be possible to relate a particular echo to a particular pulse.

Suppose we wish to have the equipment capable of measuring distances up to 185 nm. The pulse must travel 185 nm × 2 = 370 nm before the next pulse can be sent out. (PRP of time 'x'.)

$$\text{PRP} = \frac{\text{Range} \times 2}{\text{Speed}}$$

$$= \frac{185 \times 2}{\frac{1}{6}}$$

$$= 2220\,\mu\text{s}$$

The second pulse can only go out 2220 μs after the first pulse. The number of pulses that can be transmitted in one second is given by

$$\text{PRF} = \frac{1\,000\,000}{2220} \text{ pulses}$$

$$= 455 \text{ pulses.}$$

Taking the problem in reverse, let us say that the PRF of the equipment is 1000 and it is required to find the maximum unambiguous range.

$$\text{PRP} = \frac{1}{\text{PRF}} = \frac{1\,000\,000}{1000} = 1000\,\mu\text{s}$$

Now we need to calculate distance covered by 1000 μs.

$$\text{Range} = \frac{\text{PRP} \times \text{Speed}}{2}$$

$$= \frac{1000 \times \frac{1}{6}}{2}\text{nm}$$

$$= 83\,\text{nm}$$

In the above problems, 455 pulses gave us the range of 185 nm whereas a PRF of 1000 reduced the unambiguous range to 83 nm. Thus, an increase in PRF results in a decrease in the operational range.

(4) *Pulse width.* The pulse width decides the *minimum range* of the equipment. Radio waves travel 300 m in one microsecond. Therefore, for an example, a pulse one microsecond wide would extend that distance along the line of propagation. If an object at a distance of 150 m was reflecting the pulse, it would arrive in the receiver at the instant that the tail end of the same pulse was leaving the transmitter. Any object closer than 150 m reflecting the pulse will not be received as the transmitter would still be transmitting. Further, two objects in line and 150 m or less apart will appear as a single echo. Thus, if short range operation is required, for resolution and accuracy short pulses are employed, e.g. 0.1 microsecond. Larger pulses are generally employed on long range work as they carry relatively more energy in them. In practice, 1 or 2 microsecond pulses are used in medium range radar and about 5 microsecond ones for long range work.

(5) *Aircraft height.* Radar waves in the frequency bands we are discussing travel in straight lines. Because of the curvature of the earth, a considerable proportion of the surface will remain in the shadow no matter what maximum range is possible at height. The VHF formula given earlier gives an approximation of the expected ranges for given heights.

(6) *Elevation of radar head.* As you will recall, this is another factor in the VHF formula.

(7) *Precipitation and cloud returns.* At wavelengths of 3 cm and below one cannot neglect the absorption and scattering of radiation by droplets of water in clouds and falling precipitation. Cloud returns can have a most damaging effect on the performance of 10 cm and 3 cm radars and the scatter from raindrops and other weather can clutter up the display areas. Various suppression devices (e.g. circular polarisation having a rotating field) are now available to reduce rain clutter, but most work at the expense of power or range.

(8) *Intervening high ground.* If there is no clear line of sight between the transmitter and the target, radar signals will be stopped by the intervening object and that will limit the effective range in that particular direction.

Other factors, such as receiver sensitivity, bandwidth used, aerial gain in the direction of propagation, also affect the range but these are mainly equipment design considerations.

Basic elements of primary radar

Of the main components of basic radar, the master timer or trigger unit is the brain of the equipment. Its function is to trigger off a series of short electrical pulses at regular intervals. These pulses are delivered to the modulator (Fig. 9.3) and at the same time, the time base unit is advised to start timing.

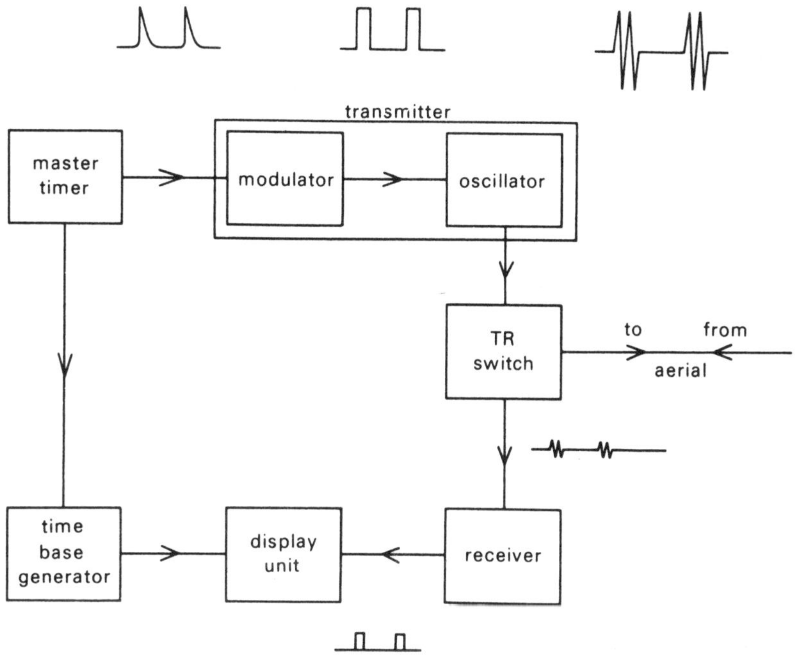

Fig. 9.3 Components of a primary radar.

The modulator's task is to generate pulses of predetermined width and sharpness. The very high voltage content of these pulses triggers off the oscillator working at the radar frequency. It is so arranged that the beginning of each pulse switches on the oscillator and the end of each pulse switches it off. Thus, the modulator acts as an on/off switch for the oscillator. The oscillator, in its turn, generates pulses of high power but short duration. The output of the oscillator is fed to the aerial.

Normally a single aerial is employed to act as transmitter and receiver aerial. A TR (transmit–receive) switch isolates the aerial from the receiver when transmission is taking place; and when the transmission is complete it switches on to the receiver to receive the pulses. Thus when the echoes are received in the aerial they are delivered to the receiver unit which, after appropriate treatment presents them to the display unit. Here they are displayed on a trace which commenced at the start of the operation under instruction from the master timer.

Advantages of primary radar

- It is a self-contained system, requiring no external assistance
- Peak power of the transmitter can be made very high owing to the relatively short time of actual transmission, but more power is required
- A common aerial may be used for both the transmission and the reception
- For a ground installation, sharp accurate narrow beams can be produced by increasing the aerial size.

Secondary radar

In this system, a transmitter (called interrogator) transmits a group of pulses on a given carrier frequency. The transmission is either omnidirectional (DME) or directed towards an object (scanner sweep of SSR). An aerial in the path of these pulses receives the signals and passes them on to the receiver. If the signals are recognised at the receiver, it instructs its transmitter (called transponder) to give a reply. The reply then goes out on a different carrier frequency.

The differences between primary and secondary radar are as follows:

- Unlike primary radar, the operation of secondary radar depends on the active co-operation of the other object
- In secondary radar the information is exchanged in the form of groups of pulses and not by individual pulses
- A secondary radar system requires a transmitter and a receiver on different frequencies, both in the aircraft and on the ground.

Advantages of secondary radar

There are various advantages of using a secondary radar over primary radar.

(1) An important advantage is the power requirement. With this type of radar it is possible to work with much lower power. There are two reasons: (a) the signals are only doing a one-way journey, and (b) there is no double scattering to combat, as with primary radar.
(2) Because of the use of different frequencies, the ground transmitter will not pick up ground reflections on transmission frequency. Similarly, the airborne transmitter will not pick up its own ground reflection.
(3) Interference through weather is reduced (see chapter 10).
(4) The system is independent of such considerations as reflecting area, shape, material, etc.
(5) In addition to range and bearing, additional information can be transmitted in the form of coded pulses.

Uses of secondary radar

DME works on the secondary radar principle. ATC uses SSR in a variety of ways. Both these systems are covered in other chapters.

Doppler radar

This topic is fully dealt with in a separate chapter.

Continuous wave radar

As the title suggests, in this type of radar, both the transmission and the reception take place continuously. Consequently, two aerials are used: one for transmission and one for reception. These aerials must be suitably screened from each other, otherwise the receiving aerial will receive signals direct from the neighbouring transmitting aerial.

Unmodulated CW may be used in the Doppler role. In this case, the transmitted and reflected signals will differ in frequency due to Doppler effect and will produce the airborne transmitter's velocity. For range measurement, the transmitted carrier is progressively frequency-modulated. The received frequency is then compared with the frequency actually being transmitted at the instant of reception, and knowing the rate of change of frequency, the range is worked out. The radio altimeter uses this principle.

The CW technique is eminently suitable for short range work. Unlike pulse radar where the minimum range is controlled by the pulse width, CW radar can work from zero range upwards. Speed-detecting radar used by the police works on CW.

Advantages of CW radar

- Because in modulated or unmodulated form, the receiver operates on a frequency different from the transmission frequency, it is fairly free from ground clutter and other permanent echoes
- It has no minimum range limitations, in altitude or azimuth
- The system is less complex than pulse radar systems.

Cathode ray tube (CRT)

The purpose of the CRT is to display visually the radar signals that are reflected by the objects. Further, by incorporating a time base, distance and bearing (or other data for which the CRT is being used) can be determined. The CRTs are classified according to the way in which focussing and deflection are achieved. There are three such classes: electrostatic CRT having electrostatic focussing and deflection devices, electromagnetic CRT having electromagnetic focussing and deflection, and lastly a combined CRT which has electrostatic focussing and electromagnetic deflection. For the purpose of the present study the electrostatic CRT is described below.

The main components of a CRT are: a cathode, a grid, three anodes and two pairs of deflecting plates (Fig. 9.4).

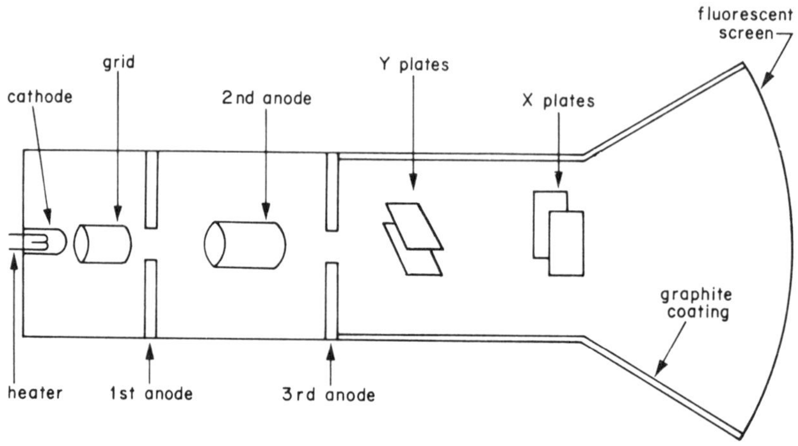

Fig. 9.4 Components of the CRT.

Cathode

The cathode consists of a small cylinder, one end of which is coated with a small quantity of barium or other similar oxide. The cylinder covers a low voltage heater which heats the barium oxide. Barium oxide when heated emits electrons.

Grid

The grid is a metal cylinder and surrounds the cathode. Its purpose is to catch as many electrons as possible emitting from the cathode and direct them in a narrow beam towards the anodes. This is done by applying a potential (called grid bias) which is negative with respect to the potential of the cathode. Electrons are negative charges and when they find that the walls of the grid are more negative than they are, they tend to be repelled from the wall and pass through the grid in a narrow beam at the centre. By varying the grid bias we can control the number of electrons passing through the grid. This is the brilliance control.

Anode system

As soon as the electrons are emitted from the cathode, the cathode becomes positive relative to the electrons and the most natural thing would be for the electrons to return to the cathode. This must be prevented; the electrons are in fact encouraged to travel forward to the screen by means of three anodes. First and third anodes have the shapes of plates while the second anode has the shape of a cylinder (Fig. 9.5).

The first and third anodes are positive, the second anode is negative. The first anode attracts the electrons which pass through its centre and then start diverging. This tendency to diverge is checked at the second anode (being

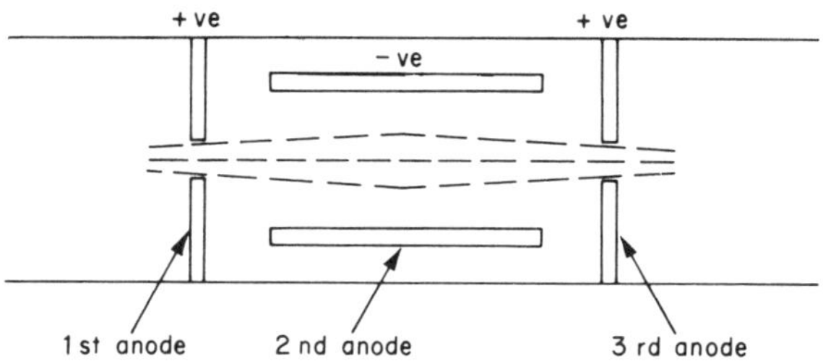

Fig. 9.5 Electron beam and anodes.

negative) and the electrons deflect back and pass through the third anode under the attraction of positive potential. When they hit the fluorescent-coated screen they show up as a glow. How sharp the glow is depends on how much divergence took place at the second anode. The potential of the second central anode can be varied to give different sharpness – this is the focussing control. As to where on the screen the electrons will hit depends on the potential of the X and Y plates.

X and Y plates

The set of plates nearest to the third anode is called the Y plates. As the electrons pass through the pair of Y plates, if, say, the top plate is positive and the bottom plate is negative, the beam will be deflected upward towards the top of the plates. This means that the beam will hit at the top of the tube, the Y axis. If we had the bottom plate positive initially and varied the potential gradually until the top plate became positive, the beam hitting the screen during this time would appear to move from the bottom of the tube towards the top. If the potential was varied quickly enough we would only be able to see a continuous vertical trace.

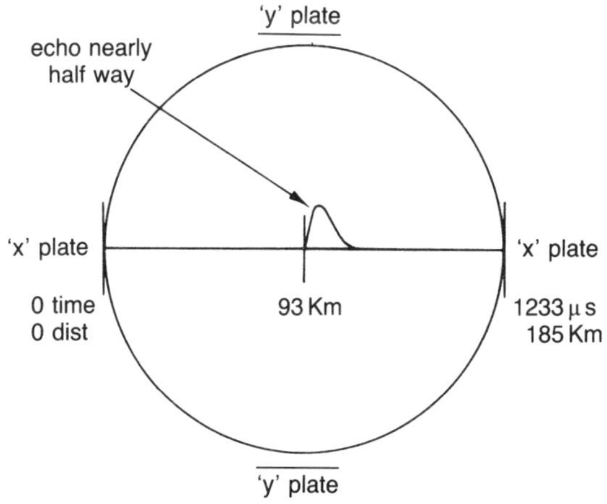

Fig. 9.6 CRT time base.

Similarly, X plates produce a trace in the X axis or horizontally. These traces are the basis for forming time bases to measure distances. For example, earlier in the chapter we calculated that a radio wave will travel 370 km in 1233 μs. If we move the spot on the CRT so that it takes 1233 μs to travel from one side of the tube to the other, what we have done is to produce a scale along which the distance of the echo could be measured (Fig. 9.6).

In Fig. 9.6 the time base is produced by the X plates, and the echo is presented through the Y plates. The time base so produced may be calibrated to read in terms of microseconds or distance, knowing that the distance from one side of the tube to another in this case is 185 km. The time base may then be calibrated by pips at convenient distances. Further, for more accurate reading, a small portion of the time base where the signal appears may be exploded to a larger scale. This is done by what is called 'strobing' the signal.

It will be appreciated that in order to produce a linear time base, the voltage (that is, potential) to the plates must be varied progressively and systematically. The voltage that has this effect is called saw-tooth voltage or waveform. This is shown in Fig. 9.7.

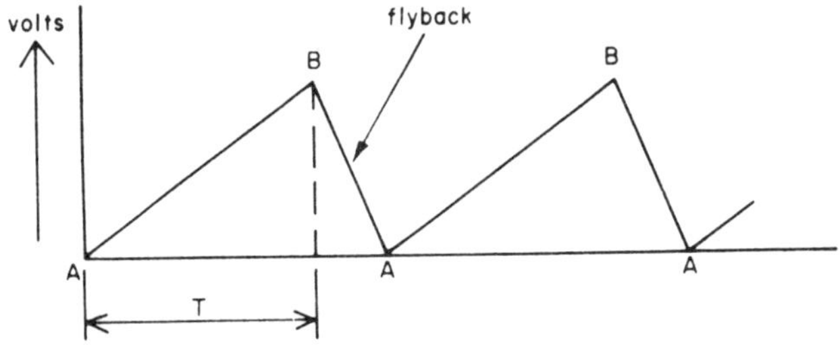

Fig. 9.7 Sawtooth voltage.

In Fig. 9.7, the voltage starting at A is increased progressively and steadily to the value of B in time T. In this example, the value of T is 1233 μs. The voltage AB is called the sweep voltage and the trace is visible. Once the beam reaches the opposite end, it must be brought back to the other end to start a new cycle. Certain time is lost as the voltage falls back to the original value. This part of the voltage change is called the flyback voltage, and it is, in general, not visible.

In practice, radar transmitters incorporate a master circuit. As the master instructs the transmitter to transmit, it simultaneously triggers the time base and the cycle commences. So the time base and echo remain in synchronism. Another method of displaying the reflected signal (echo) is to apply it as a positive voltage to the grid, to produce a 'bright-up' on the time base.

Gain control

Atmospherics and noises set up by electrical disturbances within the receiver or caused by nearby equipment manage to get on to the radar screen. These tiny signals travel on to the CRT via the Y plates and show up on the screen as multitudes of small blips in the vertical axis. These blips are known as 'grass' because of their appearance. Their presence is an essential check that the CRT is serviceable to a stage beyond producing the time base. The size of the required signal (i.e. echo) and grass is controlled by the gain control.

Cockpit displays in colour

With the rapid advances for both domestic and commercial uses of colour televisions during the 1970s, it was not surprising that by the end of the decade flight-decks would also be enjoying the facility. Now in the 1990s, for both large, wide-body commercial aircraft and much smaller, executive class aircraft, there are available – often in a single-box presentation – bright, sharp colour displays of radar information on which can be superimposed additional data, readily-interpretable symbolic displays, cautions and warnings. Many of the equipments described in the following chapters are thus available on the flight-deck with either a black and white or a colour presentation, e.g. weather radar.

Problems

(1) What is the maximum unambiguous range of a radar having a PRR (PRF) of 380 pulses per second?

Answer: 219 nm.

(2) What is the maximum PRR which can be used in a radar operating up to a range of 200 nm?

Answer: 417 pulses per second.

(3) Calculate the maximum PRF which can be used in a radar with a maximum range of 180 nm.

Answer: 463 pulses per second.

Chapter 10
DISTANCE MEASURING EQUIPMENT (DME): AREA NAVIGATION (RNAV)

DME is a secondary radar system which provides accurate and continuous indications in the cockpit of the slant distance between an aircraft and the ground transmitter. The use of primary radar is unsuitable for DME operation for a variety of reasons. The power requirements for a secondary radar are relatively low. For example, in order to cover a range of 200 nm (present limit of our DMEs) 1.5 kilowatt power output of pulses would be considered adequate (peak power output of modern equipment varies from 1 kW to 2.5 kW) using secondary radar. With primary radar, a power output of the order 1.5 megawatt would be necessary to cover a similar area.

A primary radar scanner sweeping through 360° produces a picture of the whole surface beneath the aircraft, rather than giving a range from a designated reporting point. With this radar it is impossible to create a system of positive reference points on the ground, say along an airway. Further when flying over certain types of territory (e.g. flat lands, mountainous territory), primary radar echoes may not be identified whereas with secondary radar, the ground stations can and do identify themselves.

The basic airborne system consists of:

- an interrogator (a combined receiver and transmitter)
- an indicator and
- an omnidirectional blade aerial, able to pick up vertically polarised signals.

Principle of operation

The system on the ground is called the transponder, a concocted name to describe that it is a transmitter which responds. It consists of a receiver and a transmitter. The aircraft interrogator interrogates the transponder on a given carrier frequency by sending out a continuous series of pulses in pairs. The distance between two pulses of a pair is 12 μs and the time interval between the pairs is varied at random – a technique called transmission at random PRF. At the same time that the interrogation goes out, the aircraft's receiver starts timing and commences a search for the transponder's replies.

The transponder replies to the interrogation by sending out pairs of pulses on a carrier frequency 63 MHz removed from the interrogation frequency. The receiver receives all the responses that the transponder is sending out to different aircraft but only accepts those responses which match its own PRF.

The receiver searches the responses through the maximum range of 200 nm in a matter of a few seconds (four or five seconds in newer models, 25–30 seconds with older models). During this time the pointer or counters on the indicator revolve rapidly. If no response is achieved by the time the search reaches the maximum range, the pointer (or counters) swiftly return to zero range and the search starts again. Once the response is found the receiver locks on to it and tracking commences. This is the condition which exists when the interrogator has acquired replies in response to its own interrogations and is continually displaying the slant range distance to the ground station. This distance is computed from the knowledge of the speed of the radio waves and the time taken for the pulses to travel out and back.

During the search period, the interrogator transmits at a high rate (150 pulses per second, (pps)) to achieve a quick lock-on condition. But if the lock-on is not acquired after 15 000 pairs of pulses have been transmitted, the PRF is lowered to 60 pps and maintained at this rate until the search is successfully completed. The system then operates on a random PRF between 25 and 30 pps.

Illustration of random PRF technique

Random PRF was mentioned above. This random variation in time between successive pairs of interrogation pulses prevents locking on to responses meant for some other aircraft. We will now take a closer look and see how it is done.

To keep the arithmetic simple, let us say that our equipment's PRF is 25 pps. This gives us the pulse recurrence period of

$$\frac{1\,000\,000}{25} = 40\,000\,\mu s$$

This means that if this was the PRF of a primary radar, one pulse after another would be despatched exactly at 40 000 microsecond intervals. With the DME, this time interval is intentionally varied, and it is a random variation. A pulse may be sent out 39 956 μs behind the previous one, and it may be followed by another pulse at 40 115 μs distance. The transmission pattern would look like Fig. 10.1. In the meantime the transponder is reply-

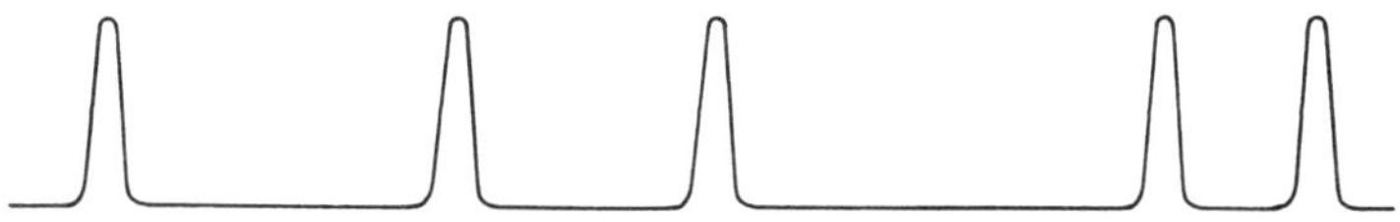

Fig. 10.1 Interrogation pattern.

ing to all aircraft triggering it. Since all these responses are on the same carrier frequency, they all arrive in the receiver. The transponder's transmissions arriving in the receiver would look as in Fig. 10.2. Some of these responses must belong to us and they are the ones which arrive with a regular

Fig. 10.2 Transponder responses.

delay from the interrogation pulses. A narrow gate in the receiver admits only those pulses which fall inside it, and the delay or the distance between the two pulses is the measure of the aircraft's range from the transponder. This arrangement is shown in Fig. 10.3. It will be seen in the figure how the pulses arriving after a regular time interval *t* (shaded dark for easy recognition) enter through the gate whereas the other responses are excluded. Of course it will be appreciated that this regular distance *t* is only momentary because unless the aircraft is circling round the beacon at a constant range, its range relationship with the station is changing all the time. However, on the microsecond timing scale this change is only minute but progressive. The gate is wide enough to accommodate these changes and it in fact moves along with the progressively changing time delays so as to keep its own responses in the lock. This technique is called the lock–follow technique. This movement is shown in Fig. 10.3, the last pulse arriving at *t* time delay.

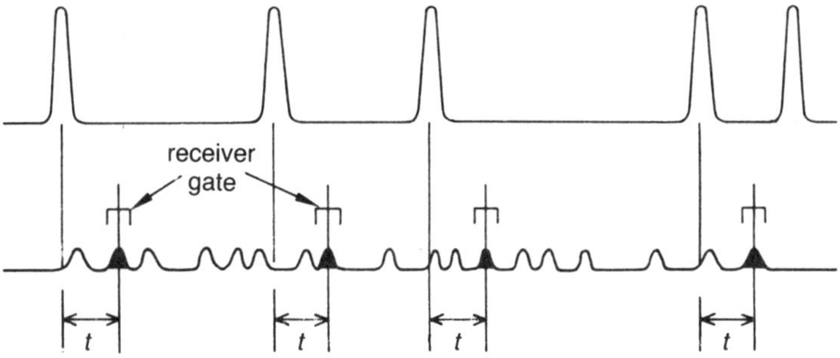

Fig. 10.3 Acceptance of own responses.

Indicators

The varieties are legion; presentation of information is either by pointer or by digital counters (Fig. 10.4). In most installations the DME is channelled by the VOR navigation frequency selection and both pieces of equipment become active together. Alternatively a separate frequency selector may be available which enables the pilot to select VOR and DME (or TACAN) as required.

The basic information is slant range from the selected station up to a distance of 199 nm. To this, an additional small computer can add the luxury of rate of change of distance display, indicate instantaneous GS, give time in minutes to the station and so forth. The advent of the course-line computer

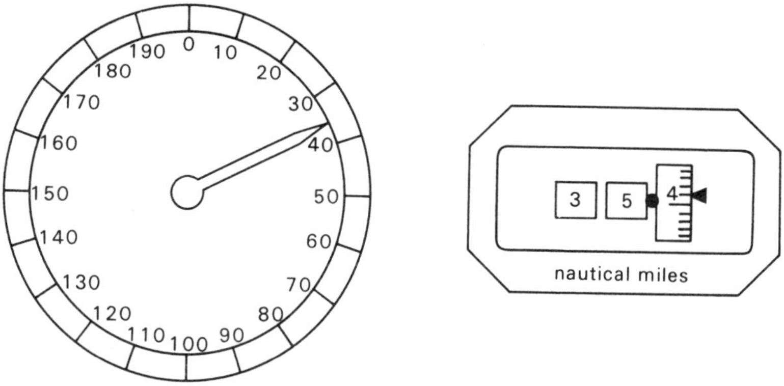

Fig. 10.4 Typical DME indicators.

opened up the prospect of area navigation. The indications are in the form of L/R deviation from your track and distance to your destination. This is in spite of the fact that your destination is not the VOR/DME station you are tuned to.

Failure indications

If the transponder reply detected by the aircraft is below a pre-set value the equipment will go on 'memory mode' for a period of around eight to ten seconds (depending on the equipment) and continue to indicate the ranges based on the last known change of range. If no signals of acceptable strength are received after this, the equipment will unlock and commence a fresh search. It will only lock on when correct signals of sufficient strength are again detected.

The unlock condition will be indicated to the pilot by an OFF warning flag on the rotary types of indicators and a bar falling across the face of the digital types. In addition the needle of the rotary indicator will rotate continuously and the numbers on the digital indicator will run up to the maximum value. Failure indication will be displayed when the equipment is switched on and:

(1) no signal is being received
(2) the received signals are below the minimum strength (just entering the DME coverage), or
(3) the aircraft is out of range of the transponder.

And of course the flag is in view when the equipment is not switched on.

Frequency and channel spacing

As we noted earlier, DME using secondary radar technique transmits and receives on different frequencies. This is a matter of necessity, because if both transmitters operated on the same frequency, assuming that this was possible and that all aircraft radiated individually coded transmissions, we would have confusion and chaos in our hands, since:

(1) the transponder's response, while arriving at the aircraft receiver, will be swamped by the ground reflections of the original transmission.
(2) the transponder's transmission will be reflected by the objects and obstacles in the vicinity of the transmitter and some of these will arrive back at the transponder. The transponder, not being able to discern between these ground reflections and the aircraft interrogations (being the same frequency), will start dishing out ranges to the reflecting objects as well. This process is called self-triggering.

With the use of different frequencies, the airborne receiver will not accept its own reflections and the ground transponder will not be activated by its transmission frequency.

DME operates in the UHF (1000 MHz) band in the frequency range of 962 MHz to 1212 MHz. The frequency allocation is divided into two bands, low and high, as follows:

Low: Aircraft transmits from 1025 to 1087 MHz; ground replies from 962 to 1024 MHz (at −63 MHz difference).

High: Aircraft transmits from 1088 to 1150 MHz; ground replies from 1151 to 1213 MHz (at +63 MHz difference).

For example, for an interrogation frequency of 1100 MHz, the response will come on 1163 MHz. Under this arrangement 63 channels are formed in each band. Channels in the low band are numbered from 1 to 63 and those in the high band from 64 to 126. These 126 channels are collectively called X channels. There is a provision for expansion into another 126 channels, to be called Y channels. We will then have, for example, a 22X channel and a 22Y channel. The channels 1 to 16 are reserved for national allocation and channels 17 to 56 are paired with VOR/ILS frequencies, as shown in the following illustration:

20X frequency paired with 108.3; 21X frequency paired with 108.4.
20Y frequency paired with 108.35; 21Y frequency paired with 108.45.

Range and coverage

DME is a short-range navigation aid providing a maximum coverage of 200 nm at 30 000 ft. The ranges indicated are slant ranges and the conversion to ground distances is by use of Pythagoras.

Example 1: An aircraft at 40 000 ft reads a DME distance of 80 nm to the station. What is its ground distance from the station?

As the distance required is in nautical miles, we must convert 40 000 ft into nautical miles before applying it to the Pythagoras formula.

$$
\begin{aligned}
(\text{Ground distance})^2 &= (\text{slant range})^2 - \text{ht}^2 \\
&= 80^2 - \left(\frac{40\,000}{6080}\right)^2 \\
&= 80^2 - 6.58^2 \text{ (approx)} \\
&= 6400 - 43 \text{ (approx)} \\
&= 6357
\end{aligned}
$$

$$
\begin{aligned}
\text{ground distance} &= \surd(6357) \\
&= 79.7\,\text{nm}
\end{aligned}
$$

Example 2: An aircraft at 24 320 ft is 30 nm ground range from the station. What is its indicated range? See Fig. 10.5.

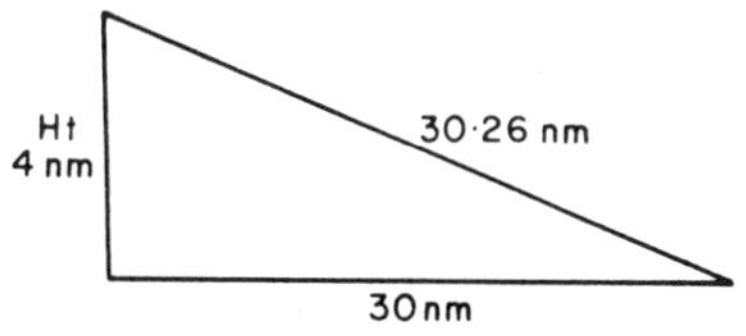

Fig. 10.5 Ground range from slant range and height.

$$
\begin{aligned}
4^2 + 30^2 &= (\text{slant range})^2 \\
916 &= (\text{slant range})^2 \\
\text{indicated (slant) range} &= \surd 916 \\
&= 30.26\,\text{nm}
\end{aligned}
$$

From the above two examples it will be noticed that the slant range errors at long distances are practically negligible. But the inaccuracy does exist which is revealed at closer range and higher altitudes.

When directly overhead the beacon, the DME will indicate the aircraft height in nm above the beacon and not zero range. For example an aircraft overflying a beacon at a height of 30 400 ft will indicate

$$\frac{30\,400}{6080} = 5\,\text{nm}$$

(When overhead, there is a small cone of silence but the range indications will continue to operate on memory.)

The actual ranges available depend on

- the aircraft height
- the transmitter height
- any intervening high ground; this will cut off the signals and reduce the range in that direction.

The ranges for various heights are worked out using the VHF formula.

$$\text{Range} = 1.25\surd H_R + 1.25\surd H_T$$

We are familiar with this formula but we will give you one more example of its employment.

Example: Give the approximate theoretical maximum range that an aircraft at 26 000 ft may expect if the DME transponder is 81 ft amsl.

$$\begin{aligned} \text{Range} &= 1.25\sqrt{(26\,000)} + 1.25\sqrt{(81)}\,\text{nm} \\ &= (1.25 \times 161) + (1.25 \times 9)\,\text{nm} \\ &= 200 + 11.25\,\text{nm} \\ &= 211.25\,\text{nm} \end{aligned}$$

Accuracy of the equipment

The receiver computes the elapsed time between transmission of the interrogating signals and the receipt of the reply signals, and determines the distance. The accuracy is of very high order. Between a slant range of 0 and 200 nm the total system error is designed to be no greater than $\pm\frac{1}{2}$ nm or ±3% of the distance measured, whichever is greater. Thus, the worst case is 6 nm at a range of 200 nm. In practice a modern DME system is considered to be inherently capable of providing an indicated range accuracy equal to ±0.2 nm or 0.25% of the slant range measured, whichever is greater. This is $\frac{1}{2}$ nm in the worst case and the figures are valid on 95% of the occasions.

Uses of DME

(1) It provides a circular position line when a single DME is used: fixes are obtained when it is used in conjunction with VOR or other DME stations
(2) Its range indication is very useful when carrying out an instrument approach
(3) It eases the task of the ATC in identifying for radar when an aircraft reports its position in terms of range and bearing from a VOR/DME station
(4) When two aircraft are using DME and flying on the same track, the positive ranges from these aircraft enables the ATC to maintain accurate separation
(5) Accurate ranges to touchdown are read off when a transponder is operating in conjunction with ILS
(6) It provides a basis for more accurate holding patterns
(7) With an additional computer, area navigation may be carried out with accuracy.

Advantages of DME as secondary radar

It will be appreciated from the earlier chapter on radar that DME uses the principle of secondary radar. The advantages of secondary radar are as follows:

(1) Interference due to weather is reduced, as seen in Fig. 10.6 below.

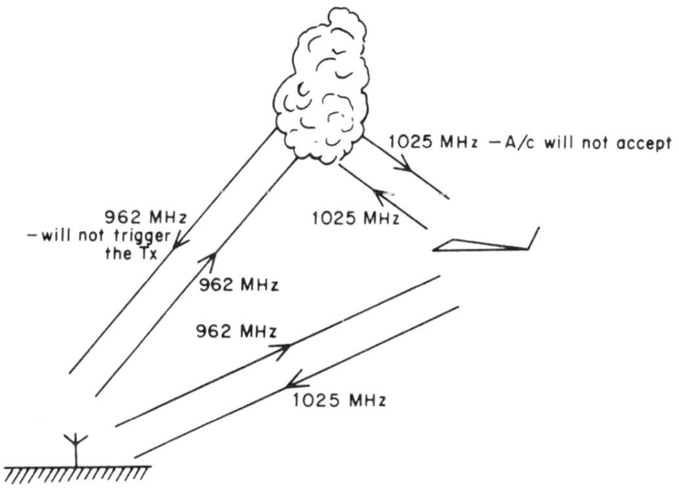

Fig. 10.6 Weather avoidance.

It will be seen in Fig. 10.6 that an active cloud not directly in line between the aircraft and the ground beacon will have little effect in causing interference or clutter.

(2) Transmission power required is only that which is sufficient to carry the signal up to the station. In other words, the signals need not be strong enough to survive a two-way journey.

(3) The ground beacon uses a different frequency from that used by the aircraft and therefore self-triggering will not occur.

Beacon saturation

Like a shopkeeper who opens the doors of his shop in the morning and makes available any and all of his wares to the customers, once he is sold out, he puts the shutters down; the ground equipment, when switched on, does likewise. The transponder transmits 2700 pps at random whether or not it is being triggered for information. These pulses are available to its customers. When an aircraft interrogates the transponder it replies by using some of these random pulses. Now, unless an aircraft is in search mode, its normal operating PRF is 25–30 pps, average 27. One aircraft triggering a transponder and in lock-on condition replaces 27 transmitter random pulses. At this rate if 100 aircraft are simultaneously triggering a station, the transponder's capacity will be exhausted and the beacon would become saturated. In arriving at this figure we did not consider those aircraft which are operating on higher PRF in the search mode. They would put excessive loading on the transponder's capability. However, the search mode at higher PRF runs for such a short time that it can be discarded for practical purposes, and the search at 60 PRF is not normally continued for a long time because if it is not locking, it is very likely that you are still out of range.

When the beacon becomes saturated it adjusts itself to cope with the situation by reducing receiver gain, Fig. 10.7. If you are receiving strong

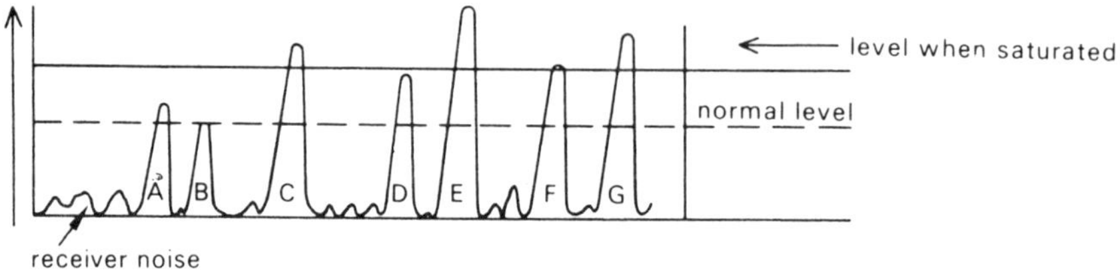

Fig. 10.7 Beacon saturation.

music on your transistor radio which is accompanied by weak background noise, you turn the volume down. In doing so, you aim to exclude the weaker, unwanted noise. The transponder, by reducing its receiver gain excludes interrogations from aircraft whose pulses reaching the transponder are relatively weaker. In Fig. 10.7, in normal operation all aircraft from A to G would be receiving ranges from the transponder (aircraft B is just entering the coverage and might start receiving soon). When the beacon gets saturated and the receiver gain is reduced, aircrafts A, B, D, and possibly F will be excluded from service and an unlock will occur in the airborne equipment. The purpose of the system is to give preference to the nearest aircraft but this is not necessarily achieved as it responds to the 100 strongest signals. The UK is commissioning a new generation of DME transponders with higher power (ranges much greater than 200 nm) and greater capacity (considerably more than 100 aircraft).

VOR/DME planning

A VOR provides magnetic bearing information. A DME provides slant ranges from the station. When these two equipments are used together, we can have instantaneous fixes in the form of a bearing position line and a circular range position line. Further, if the two transmitters are co-located, these two position lines may be plotted from a single point.

To achieve speed in selecting the facilities in the cockpit and to reduce the workload on the flight crew, VORs may be frequency paired with DME or TACAN (military installations) stations. This means that when a VOR frequency is selected, the DME circuits would be activated automatically. Ideally, VOR and DME meant for use in conjunction with each other should be co-located, that is, both the transmissions made from the same geographical point. However, this is not always possible and where VOR and DME stations are not widely separated they may still be used in conjunction with each other. The pilot will know the relationship between the two stations by noting the ident signals and frequency pairing arrangements as explained below.

(1) Where both VOR and DME/TACAN transmit the same callsign and in synchronism, the stations are called associated and they are always frequency paired. The term 'associated' means that
  (a) the two transmitters are co-located (i.e. the two antennae are co-axial)
  (b) they are a maximum distance of 100 ft apart where the facilities are used in the terminal areas for approach purposes, or
  (c) they are at a maximum distance of 2000 ft apart where their purpose is other than (b) above but where the highest position fixing accuracy is required.

Synchronised idents are transmitted every $7\frac{1}{2}$ seconds, that is, each 30-second period is divided into four equal parts. The DME transponder transmits its ident during one of these four periods, and the VOR in the other three.

(2) Those VOR and DME stations which are not associated but serve the same area (approx 7 nm) and which may be used in conjunction with each other are also frequency paired. But in this case, both VOR and DME will identify separately and one of the two will have a letter Z in the callsign, e.g. STN–STZ.
(3) Where VOR and DME stations are at entirely different locations, they may or may not be frequency paired. Both facilities will have independent idents. Note that when a VOR is frequency paired with military TACAN the system is called VORTAC.

DME ident is made up of a series of paired pulses at PRF of 1350 pps. A decoder in the receiver converts the information and feeds it into the earphone as morse letters.

Miscellaneous

The type of emission is 'PON'. A typical UKAIP entry records transmission and reception frequencies, e.g. transmit 1174, receive 1111. These apply at the station and mean that the aircraft transmits at 1111 and receives at 1174. In any case you select a channel number which in this case is 87X.

Area navigation (RNAV)

In the CAA (ECAC) airline transport pilot licence syllabus, immediately following the entry DME comes VOR/DME Area Navigation (RNAV). Although navigation and planning appear in their own right as syllabus subjects (and also separately in *Ground Studies for Pilots* volume 2), it is also appropriate that RNAV principles and operations should appear in the sequence of radio aids topics, it being so dependent on VOR and DME.

Principle of operation

The present route Air Traffic Service (ATS) structure, which is particularly complex in the European/Mediterranean region, evolved from 'point-source' aids between which aircraft are required to fly. Initially there were radio ranges then NDBs followed by VORs and DMEs. This structure offers little scope for expanding traffic capacity, the options open to pilots or to improvement on present levels of efficiency which already impose high workloads on air traffic controllers and pilots, especially in terminal airport areas. It is believed that development of RNAV throughout the 1990s will overcome the present deficiencies and enable ATS systems to accommodate the increasing need for operators to enjoy a greater route flexibility and traffic capacity, handled safely and efficiently.

In ICAO Annex 11, RNAV is defined as a method of navigation which permits aircraft operation on any desired flight path within station-referenced navigation aids, or within the limits of the capability of self-contained aids, or a combination of these. Thus, in general terms, RNAV may be considered as any system of navigation which is capable maintaining track and time to a specified degree of accuracy *without having to overfly a point-source aid.* This can be done by simultaneously fixing position with VOR/DME or DME/DME etc.

Advantages and disadvantages

Unfortunately RNAV cannot be immediately implemented because aircraft using airways vary considerably in their age, their performance and the degree of sophistication of their on-board equipment. This leads to varying standards of navigation even though the flight crews are meeting the internationally-agreed mandatory minimum requirements for airways flying. Even the national ATS systems in adjoining countries may differ in operating concepts and procedures, although they are within the international parameters. The advantages of adopting the RNAV principle over the current fixed-route system stem from the introduction which will then be possible of more direct routeing of aircraft so reducing flight distances, times and fuel required. It could also lead to increasing existing, or new, route capacities by enabling the use of dual or parallel routes, reduced separation horizontally and vertically between routes and of the basic volume of protected airspace. It will allow pilots and operators to exercise greater freedom of choice while also giving ATC greater flexibility.

Accuracy, reliability and coverage

There must inevitably be a transitional stage with the introduction of RNAV because there will still be aircraft operating wholly dependently on overflying point-source aids. In Europe, time scales are being established through the 1990s and initially RNAV will be used within the existing ATS route system. Then it is envisaged that as most aircraft come to meet the minimum equipment standards, there will be the following types of RNAV routes:

Fixed RNAV routes: These will be published permanent ATS routes which can only be flight-planned by aircraft with the approved RNAV capability.

Contingency RNAV routes: These will be published ATS routes usable by aircraft with RNAV capability during specific time-limited periods.

Random RNAV routes: These will be unpublished routes which can be flight-planned within certain designated RNAV areas.

Within the RNAV concept itself, there are two recognised levels of accuracy of operation: B-RNAV and P-RNAV. Basic RNAV (B-RNAV) has an accuracy comparable with that of aircraft currently operating the present system on routes defined by VOR/DME. Precision RNAV (P-RNAV) requires a track-keeping of 0.5 nm standard deviation or better.

Presentation, interpretation, FMS

Already many UK-registered aircraft are fitted with RNAV capability equipment and as the phases of introduction are implemented so, in the UK, the Air Navigation Order will be amended to lay down the rules for the approval of RNAV equipment. It will also lay down the installation and maintenance rules for equipment together with the operational procedures to be used. The current position is laid down in Articles 39, 39A, 39B of the ANO and expanded in a yellow (Ops/ATS) AIC. The intention is that carriage of RNAV equipment will be mandatory from 1 January 1998 within the airspace of ECAC member states.

The interpretation of the individual instruments, VOR, DME, ADF, etc., has been described earlier when considering the basic systems. In companion volume 3, the FMS to be found on the flight deck of modern airliners is described together with its use. With the availability of the on-board computer and predetermined waypoints, pilots will be able to fly any fixed, contingency or random RNAV route stored in the FMS memory in the same way as the airways are currently flown between the fixed-point aids.

Test questions

(1) Describe the basic principles of DME operation.

(2) Explain why for DME the ground transmitter operates on a different frequency from that transmitted from the aircraft.

(3) Explain how it is possible for a number of aircraft to use the same DME beacon simultaneously without mutual interference.

(4) The type of emission used for DME is:
(a) A8W (b) PON (c) A9W.

(5) The maximum number of aircraft that a DME beacon can handle before becoming saturated is:
(a) 63 (b) 126 (c) 100.

(6) On a DME with digital presentation, failure indication is given by:
 (a) the DISTANCE TO GO returning to 0
 (b) a drop-down bar falling across the face of the figures
 (c) the DISTANCE TO GO oscillating at figures in excess of 200 nm.

(7) DME operates in the frequency band:
 (a) VLF (b) MF (c) UHF.

(8) The range from the beacon indicated by DME is:
 (a) slant range (b) ground range
 (c) ground range only if the beacon is co-located with VOR.

(9) If VOR and DME stations have separate identifications of 'VON' and 'VOZ' for VOR and DME respectively, this means:
 (a) the VOR and DME beacons are co-located
 (b) the VOR and DME beacons are not co-located, but are serving the same location and may be used in conjunction with each other
 (c) the VOR and DME beacons are at entirely different locations.

(10) If a DME beacon becomes saturated, it adjusts itself to:
 (a) give preference to the strongest aircraft signals
 (b) give preference to the most distant aircraft
 (c) give service to a maximum of seven aircraft, irrespective of distance.

Chapter 11
SECONDARY SURVEILLANCE RADAR (SSR)

The variety of aircraft types with wide differences of speed and altitude in a crowded airspace demands positive identification of each aircraft for adequate safe control by ATC. Primary radar is insufficiently informative, and has the added disadvantages of clutter on the screen and a necessarily high power output for the two-way journey of the wave. Secondary radar, when used in conjunction with the primary radar, does away with these drawbacks, but does demand the co-operation of the aircraft in that the appropriate equipment must be aboard. When such equipment must be carried in designated UK airspace is firmly and legally laid down.

A ground based transmitter/receiver triggers off a reply from an aircraft's receiver/transmitter when the correct operating procedures are followed: the reply is on a different frequency from the interrogator. The interrogator is the name for the ground equipment, the transponder for that in the aircraft. The aircraft not only identifies itself positively without manoeuvres, but gives its height. On the ground, range and bearing are displayed on the screen, nice and clear, while the aircraft's height, destination or point of leaving airspace and callsign are also shown on the screen as computer-generated symbols on a synthetic clutter-free display. All signals are coded; the code of the interrogation signal is called the mode.

Frequency (UHF)

Ground transmits on 1030 MHz, receives on 1090 MHz;
Aircraft transmits on 1090 MHz, receives on 1030 MHz.
The ground interrogator aerial is directional while the aircraft transponder aerial is non-directional.

Process

The method used is the transmission and reception of pulses, and it is essential to eliminate weak or spurious signals, since the coding system depends on the 'presence' or 'non-presence' of pulses. The interrogator transmits two pulses with a known spacing, and there are four modes, each mode having a different spacing.

Mode A has pulses (always 0.85 μs wide) 8 μs apart
Mode B 17 μs apart

Mode C 21 μs apart
Mode D 25 μs apart.

Modes A and B are used for identification, Mode C for automatic height information, while Mode D is experimental. The aircraft transponder will reply to an interrogation signal provided the pilot has selected the corresponding mode. The transponder transmits a code in reply to a correct interrogation (correct in that the aircraft equipment recognises the mode by the time spacing between each pair of interrogation pulses), a code which is obtained by the inclusion or omission of any of up to 12 pulses. The train of 12 pulses is contained between two framing pulses, 20.3 μs apart, and these are always sent. Between them, the information is sent by transmitting or leaving out any of the 12. The codes available in a twelve-pulse train then are $2^{12} = 4096$, and the codes are numbered 0000 to 7777, using all numbers except those containing an 8 or 9. Pulses in the transponder are 0.45 μs wide. A further pulse, the special identification pulse, can be transmitted when the ident button is pressed on the aircraft unit, usually at ATC's request; this pulse is after the second frame pulse, and will be automatically and continuously transmitted for about 20 seconds after pushing the button.

The modes and codes are selected by switches on the aircraft control box: a function switch for mode, a window for code, a button for ident, and a switch for automatic height reporting. The mode and code are pre-allotted before departure usually, or requested by ATC in flight; there are various special selections such as Mode A, code 7600 to be used in the event of radio failure, 7700 for distress, 7500 for hijack, to quote examples. Others are 0000 to indicate a transponder malfunction, 2000 for entering airspace from an area where SSR operation has not been required, 7000 for conspicuity code. 7007 is allocated to aircraft engaged on airborne observation flights under the terms of the Treaty on Open Skies (which are published in NOTAMs).

When selecting a code, this should be done on the STBY (standby) switch position. Then when flicking over the counter in the window (or on the LCD display) to the required code, there will not be any risk of inadvertingly transmitting a whole string of codes until you reach the required four-figure number. Particularly it is important not to squawk accidentally any of the emergency settings. If any emergency setting is transmitted, it immediately activates an alarm system at the ground station and automatically initiates the emergency procedures. Having made the required setting, the function switch is then set at ON, or at ALT if requested to SQUAWK CHARLIE. . . .

Automatic altitude telemetering

On getting Mode C interrogations, the transponder will produce one of 4096 codes, no matter what code is selected in the window. This code is determined by the output of an altitude digitiser mechanically linked to the altimeter; the sequence of pulses transmitted is thus entirely determined by the aircraft's height. This height is always referenced to 1013.2 mb, quite independent of altimeter setting; the equipment will provide automatic altitude telemetering up to 128 000 ft, with a change of output every 100 ft. The controller is thus automatically provided with the aircraft's flight level. If your allocated flight level is 65 and you let the aircraft wander below the cleared level, then as soon as you pass from 6450 ft to 6445 ft the controller's 'blip' coding will immediately show you flying at 6400 ft rather than your cleared 6500 ft.

Unwanted echoes

The interrogator aerial sends out a wide vertical beam and a narrow one in azimuth: the azimuth beam, though, has side lobes which could produce a transponder response, spreading the echo on the indicator tube and denying the required accuracy of range and bearing. To correct this, an omnidirectional radiation transmission is introduced, whose signal strength is greater than the strongest side lobe but less than the main beam. By fitting a circuit in the transponder for comparing the amplitude of pulses, it can be arranged not to reply to side lobe interrogations; for example, the first pulse of the mode can be transmitted in the omnidirectional pattern, and the second in the interrogator pattern; the transponder will only reply if the interrogator pulse is equal to, or greater than, the amplitude of the omnidirectional pulse. Or, by a normal transmission of interrogator pulses with an omnidirectional pulse intervening 2 μs after the first: the transponder will not reply if the omnidirectional pulse is greater in amplitude than the interrogator.

Despite the circuitry, some 'fruiting' and 'garbling' may occur. If aircraft are within the range of two or more SSR stations they may cause non-synchronous interference to one of the stations by responding synchronously to another station, which is called fruiting. Also if two or more aircraft are close enough together, such as when in a holding pattern or overtaking, so that they are in the aerial beam at the same time and produce overlapping replies, this is known as synchronous garbling.

General

The aircraft equipment is kept on Standby until required; this keeps the display on the ground clean. The range of SSR is of the order of 200 nm, and the PRF is about 250 per second. Several aircraft in an area with similar flight plans may have been allotted the same code; identification of one would be demanded by ATC, and the resultant echo on the ground display would show as a 'filling-in' of one of the echoes already showing; or ATC might of course order an aircraft to turn to another code.

Advantages

- Longish range
- No clutter, no unwanted echoes from cloud, high buildings, high ground, and so on
- Reply signals give range, bearing, height, aircraft identity, destination or point of leaving airspace positively and automatically

- No effort required by the pilot – well, very little anyway
- All other communication channels are left free
- Information of ATC is instantaneous and unambiguous
- No aircraft manoeuvres required
- Little power needed.

Disadvantage

Disadvantage is that the aircraft must carry the necessary equipment.

In UK airspace, stringent regulations for the carriage of SSR transponders apply.

It is compulsory when flying in the Upper Airspace and whole of the UK controlled airspace under instrument flight rules (IFR) to carry Mode A 4096 codes and also Mode C.

Gliders are exempt from the above requirement, as are aircraft below FL 100 in controlled airspace receiving an approved crossing service.

Most ICAO States have now published some form of mandatory requirements for carriage of SSR, the details of which are generally available from the State's AIP as well as from the Aerad Supplements.

There is a standard RTF phraseology for SSR, the operative word being 'Squawk'. For example: 'Squawk Alpha Code 7600' means 'Select Mode A Code 7600 on your control box, and switch on transponder'. 'Squawk Ident' means 'Stay on present mode and code, but press the Identification button'.

Mode S data link

In chapter 3, reference was made to ground/air two-way communications by means of VHF or Satcom. A further datalink is now being introduced through what is known as SSR with selective addressing – Mode S, specified in ICAO Annex 10. Whereas the SSR modes and codes so far described handle only 4096 identities with altitudes in 100 ft increments, Mode S has enormously greater capacity, while the equipment will still be compatible with the current Modes A and C units. It will handle over 5×10^{33} different uplink and downlink messages and have a capacity for over-the-horizon service because it will also have a widely-distributed ground data network. The uplink message will be specifically addressed to a particular aircraft and the system permits for over 16 000 000 discrete aircraft addresses – far in excess of the number of aircraft of all types in existence world-wide! (see Fig. 11.1.)

Mode S is designed to serve two main functions – communications and surveillance. It is foreseen that the great bulk of RTF messages will be eventually be exchanged by datalink, whether air-initiated or ground-initiated. With FMSs becoming commonplace, messages to and from the flight crew can be managed via the CDU. From the small aircraft aspect, the proposed Mode Sierra transponder kit will, compared to the panel of the current Modes A and C transponders, also have a CDU format on which messages on weather, destination, ATC, etc., can be displayed to the pilot in flight. As regards surveillance, the encoding altimeters will have a capability to report

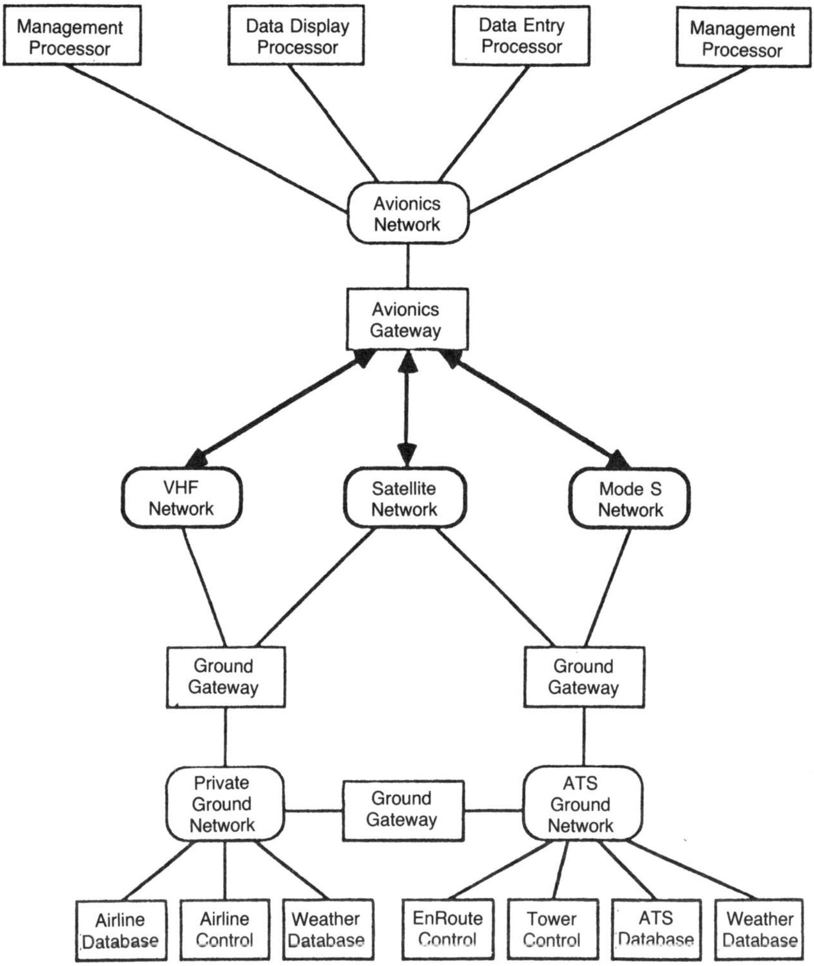

Fig. 11.1 3 'carriers' giving ground-air-ground communications

in 25 ft increments rather than the 100 ft increments of Mode C. Enhanced data in the form of the aircraft's bank angle, accelerations, heading, ground speed, etc., will all pass automatically to ATC. The feeding of the bank angle for example, will in turn enhance the radar tracking on the screen in front of the controller. Currently there is a slight delay in displaying a turn initiated by an aircraft because the ATC system computer checks that it is indeed a genuine turn on to a new heading and not just a transient 'jink' in the response. Because in normal flight the pilot applies bank when initiating a turn, the bank information can supplement the radar plot report, so telling the computer immediately the turn starts. All of the finer detail arriving from the enhanced actual data from the aircraft and the flight intentions with the completely unambiguous aircraft identity will enable better conflicting traffic information and its resolution. Airborne collision avoidance systems (ACAS) and traffic conflict alert system (TCAS) themselves are discussed in a later chapter.

Chapter 12
GROUND RADARS

Ground radars are used extensively in civil aviation, particularly by the ATC services, for which they now form an essential part of their equipment. The main types of ground radars may be summarised as:

(1) ***Long range surveillance radar.*** Used for airway surveillance, etc., up to 200–300 nm. Primary radar provides range and bearing, i.e. positional information, which in some areas may be supplemented by a height-finding surveillance radar. Additional information is also provided by SSR.
(2) ***TMA surveillance radars.*** Medium range radars, up to 75 nm, used for controlling traffic in terminal areas, etc. Primary radar provides positional information which is normally supplemented by SSR.
(3) ***Aerodrome surveillance (approach) radar.*** A short range, approximately 25 nm, primary radar providing positional information. Used for the control of aircraft in the vicinity of an aerodrome and for limited approaches.
(4) ***Precision approach radar (PAR).*** Gives very accurate azimuth and elevation guidance relative to the approach path to an aerodrome runway. Used originally by a ground controller to 'talk down' a pilot to the runway; in the UK this procedure has now been withdrawn from most civil aerodromes.
(5) ***Surface movement radar.*** Installed at major airports to provide a very precise radar picture of the aerodrome surface. It is used in poor visibility to control the movement of traffic on the aprons, taxiways and runways. Maximum range approximately 2.5 nm.
(6) ***Weather radar.*** A primary radar used by the meteorological service to supplement their knowledge of current weather conditions, varying from thunderstorm location to finding upper winds.

Note: Short-range radars, e.g. surface movement radar, employ very short wavelengths (e.g. 1 to 3 cm) to achieve short rectangular pulses which give superior resolution and accuracy. Long-wave radars, e.g. surveillance radars, employ longer wavelengths (e.g. 10 cm) with larger pulses to achieve greater ranges with less attenuation.

To enlarge upon (3) and (4) above, an approach surveillance radar (RAD) and PAR may be utilised for approach and landing down to zero conditions.

RAD

It may be necessary to ensure firm identification, to call on the aircraft to perform a simple manoeuvre (e.g. a procedure turn) and follow its blip on the radar screen. Other methods of identification acceptable are a VOR/DME fix, a response from the aircraft's SSR transponder or a direct handover from one radar unit to another. Once identified, the aircraft is guided verbally to the approach path at 1500 ft and is handed over to the precision controller. ICAO lays down a minimum specification of being able to identify a small single-engined aircraft at a distance of 20 nm at 8000 ft. The equipment must give a position accuracy of ±2° of the true position and be able to see two aircraft separately when 4° apart from each other in azimuth. This is achieved by radiating a beam 2° in azimuth to an elevation of 30°. The scanner rotates at 10, 15 or 20 rpm and the system is duplicated. The display face is switchable to scales of 10, 20, 40 or 60 nm range.

The use of a relatively long wave produces only weaker rain echoes and when working on a 50 cm (600 MHz) band it is completely clear of the rain clutter and therefore no suppression device is necessary. ASR is essentially an instrument approach system where the air traffic control officer (ATCO) issues instructions so the pilot can bring his aircraft on to the final approach path on the extended runway centreline. It is an azimuth and range facility but has no elevational data.

PAR

Precise information must be readily visible for immediate instructions to the aircraft, so two discriminators of 3 cm wavelength are available. The beam sweeps both in azimuth and in elevation along the approach line. The antennae are sited to one side of the runway near the touchdown point. The system requirements are that it should be capable of detecting and indicating the position of an aircraft of 15 m^2 (165 ft^2) echoing area (or larger) which is within a space bounded by a 20° azimuth sector and a 7° elevation sector to a distance of at least 9 nm. The maximum permissible error from on-course indication is 0.6% of the distance from antenna plus 10% of deviation from the on-course line or 30 ft whichever is greater. Similarly, elevation accuracy is 0.4% of the distance from the antenna plus 10% of the actual linear displacement or 20 ft, whichever is greater.

The information is displayed on two screens mounted one on top of the other and the controller is able to pass instructions to the pilot by watching two blips in relation to the centre lines on the screens.

General notes

It is important that when PAR is on a location also served by ILS that glidepath and azimuth indications are coincident inbound from the outer marker. Two pilots in CAVOK will practise the two approaches, one monitoring the other. In certain weather conditions it is mandatory for ILS approaches to be monitored. In this case the precision controller will advise the pilot that his approach is being monitored, but the controller will take no action as long as the pilot remains inside the ILS funnel ($\frac{1}{2}°$ above/below the glidepath and 2° either side of the centreline). The pilot will be warned if he goes outside the funnel and the PAR will take over if the pilot requests.

The procedures are precisely laid down in the UKAIP, and the pilot on the approach is on airfield QFE, azimuth and elevation information is given absolutely and in relation to the on-course and glidepath line. See Fig. 12.1.

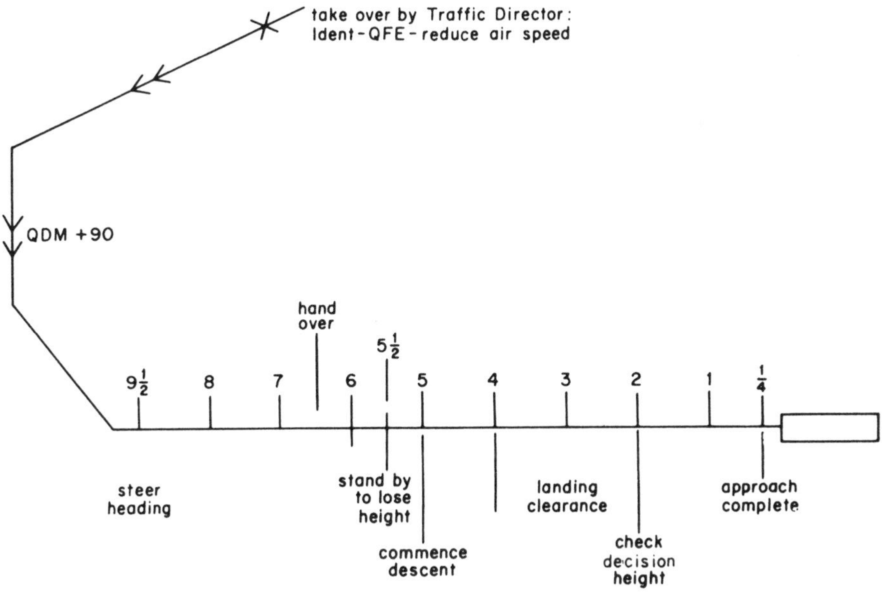

Fig. 12.1 PAR procedure.

At 7 to 6 miles, the traffic director hands over to the precision controller. Satisfactory RTF contact, then do not acknowledge further instructions. Prepare to descend, and commence descent to maintain a $x°$ glidepath. Azimuth instructions and information as to position with regard to centreline from hand-over to completion.

Elevation instructions and information as regards glidepath from start of descent to completion.

4 miles out – 'clear to land, surface wind so and so'
2 miles out – 'check decision height'
$\frac{1}{4}$ mile out – 'approach completed'

After landing contact tower on frequency . . . for taxying instructions.

It is the pilot's obligation to break off if not in visual contact at DH. The controller throughout gives definite headings (M) to fly, so he has made wind adjustments for the QDM to make good. He usually, too, gives the range every mile until four miles off, every $\frac{1}{2}$ mile thereafter – sometimes every $\frac{1}{4}$ mile after $2\frac{1}{2}$ miles off. The last range given is $\frac{1}{2}$ mile, when all the crew are searching for the approach lights.

Surveillance radar approaches

At airfields where either PAR is not available or the glidepath element of the PAR is unserviceable, approaches may be made on surveillance radar but this is without the benefit of the glidepath. The instructions passed to the pilot are similar to PAR except that in the absence of the glide slope, check heights are passed, e.g. 'your range is 4 nm, your height should be 1200 ft'. Depending on the accuracy of the surveillance radar, the approaches are terminated either when $\frac{1}{2}$nm from the threshold or 2 nm. These terminal distances are also given in the UKAIP.

The frequencies given in the UKAIP are the actual RTF frequencies and not the frequencies on which the radar operates.

Range of the surveillance radar

The range that a primary radar produces depends upon:

- transmission power
- PRF used
- type of suppressor in use.

The range at which an aircraft will be detected depends upon

- the aircraft height
- intervening high ground
- the aircraft shape, size and material
- the weather conditions.

Accuracy

Within the limitations of the radar, the accuracy of the approach depends primarily on the skill of the controller and also the ability of the pilot to rigidly follow the instructions.

Advantages/disadvantages

The advantages are clear: no special gear is required in the aircraft, the pilot has only to obey RTF instructions, no interpretation of meters is involved and the search system provides a means of partial traffic control. Above all, the system is movable from runway to runway to be operational within half an hour.

The disadvantages are: several RTF channels are needed at busy airfields, landing rate is limited, ground controllers must be highly skilled (as well as pretty durable), identification of blips can be difficult and there will be some clutter in rain or snow (remember that a pulse is reflected from objects of comparable wavelength – PAR is 3 cm).

Break off

The pilot must break off the approach at the DH unless a visual landing can be made. The controller has authority to order an overshoot.

Chapter 13
AIRBORNE WEATHER RADAR (AWR)

AWR primarily provides a pictorial representation of turbulent and dangerous clouds located along the flight path and warns the pilot well in advance. The information is displayed on a CRT and the picture can indicate the best route of penetration through bad weather. As a secondary function the equipment also provides the pilot with the facility of map-painting radar for the purpose of navigation and avoidance of high ground. On EFIS-equipped aircraft, the display appears on the navigation display either on its own or integrated with navigation information, TCAS etc.

Principle

AWR is a primary radar. In both mapping and cloud detecting roles, the requirements are to find the range and bearing of the objects. The range is found by the echo principle and the direction is found by use of the searchlight principle. Both these techniques are discussed in the chapter on radar.

Weather radar

The requirement is to detect turbulent cloud, provide information on weather severity and indicate safe routes round, if any. The efficiency of the equipment in discharging the above functions depends on the following two factors:

(1) ***Wavelength (or frequency).*** As noted earlier, the efficiency of the reflection depends on the wavelength in relation to the reflecting object. For AWR, the wavelengths considered appropriate for satisfactory operation lie between 10 cm and 3 cm in the SHF band. At these wavelengths fine mist, haze, clouds of tiny water droplets do not reflect energy whereas water droplets in a cumulonimbus will have reached sufficient dimensions to give an echo. In this way, harmless stratus is prevented from cluttering up the screen.

If the wavelength is increased above 10 cm the waves will be too large and no reflections will occur. On the other hand if it is reduced too much below 3 cm, a relatively large amount of energy will be absorbed instead of being reflected, although some radar does work on a 2 cm wavelength.

Thus, the choice being limited between 10 cm and 3 cm, for a given power output, the energy on a 3 cm wavelength will give a better range and a more comprehensive picture of the situation. On the other hand, at a shorter wavelength it will possibly not look behind a reflecting cloud. A 3.2 cm radar is quite popular with commercial transport aircraft.

(2) ***Beamwidth.*** The beamwidth must be kept as narrow as possible for good target resolution. Two targets less than a beamwidth distance apart will appear as one single target, see Fig. 13.1. The beam widens with range. Therefore, for example, two objects at 100 nm distance may appear as a single target, but as the range is closed and the beam narrowed, each will establish a separate identity. To an operator this may give a false impression of fresh activity being developed at a close range. Therefore, the narrower the beam, the better the resolution; and for airborne equipment this again means the use of shorter wavelengths.

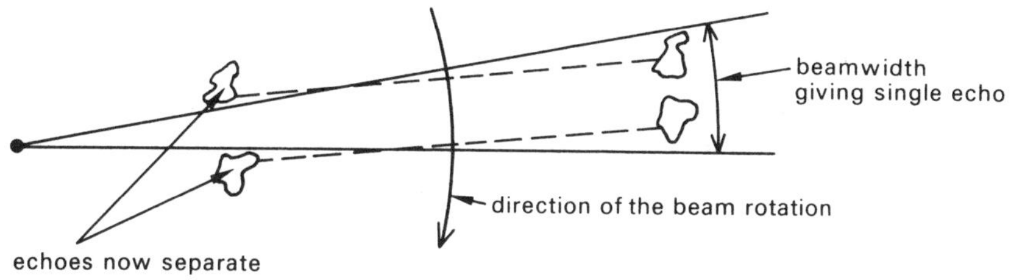

Fig. 13.1 Effect of beam width.

From the above study we can conclude that the factors which determine whether or not a cloud within the equipment's range will be detected or not are:

(1) the size of the water drops and
(2) the wavelength or frequency in use.

The accuracy of resolution will depend on the beamwidth. A conical pencil beam is considered to be the most suitable for use with AWR giving a good range for power rating.

It is because AWR is dependent upon returns from water drops, hailstones, etc., to indicate the associated turbulence that it is not a system which will display clear air turbulence such as associated with jet streams.

Map-painting radar

Mapping, again, should be done on a narrow beam in the interest of resolution (the beamwidth used is around $3\frac{1}{2}°$). It should be broad enough to cover a maximum possible horizontal distance. Such a beam would be a fan-shaped beam, see Fig. 13.2.

It will be appreciated that with this type of beam the surface illuminated below and closer to the aircraft will give stronger (and thus brighter) reflec-

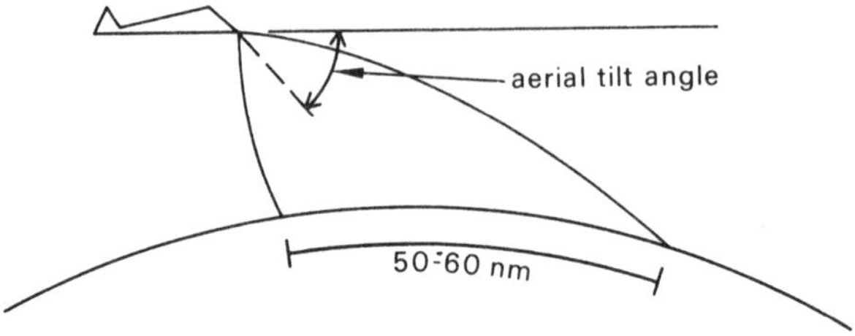

Fig. 13.2 Cosecant (mapping) beam.

tions than the ground further away. The picture on the CRT will appear progressively fading away with distance. This in itself may not be an inconvenience but radar maps are essentially read from the differing brightness of the dissimilar objects. Some objects come up very bright, others not so much, and yet others which do not reflect at all. For a reliable interpretation we would like all similar objects to produce similar brilliance. This difficulty is overcome by adjusting the power spread so that maximum power is directed to the farthest point to be covered. The power is progressively reduced as distances decrease so that the power directed to the closest object is minimum. This reduction in power with decreasing range is the function of cosecant of the depression angle and the beam so produced is called a cosecant (or mapping) beam.

Theoretically, for the maximum range the upper edge of the fan should radiate horizontally ahead of the aircraft, that is, at a depression angle of 0°. This cannot be achieved in practice as the cosecant of 0 is infinity. A practical beam is tilted down by about 5°. In flight the aerial tilt angle can be adjusted to give a most advantageous display.

As the cosecant beam is more widely spread out than the pencil beam used for weather detecting, and as the power available to both is the same, the distance of coverage available using a cosecant beam will tend to be more restricted. The actual range available depends on the power, the aircraft's altitude, the reflecting property of the surface and the angle of depression in use. With older equipment 70 nm is just about the limit. Use a pencil beam where distances beyond the cosecant beam's range are required to be scanned.

The shapes that you will see on the CRT, such as a coastline, or the contours of a hill or the outline of a town, will not be identical to what you would see with your naked eye. Figure 13.3 shows a straight coastline across track 6 nm ahead, Fig. 13.4 how the beamwidth affects the size of the echo displayed.

Little or no energy will come back from a calm sea. This makes a coastline easily separable from the landmass. Similarly fine sand or flat terrain will not give reflections. Brighter echoes will occur in built-up areas, taller buildings showing up brighter still. Hills and man-made structures cast shadows behind them, Fig. 13.5, sometimes giving a false impression of water or a lake. If you

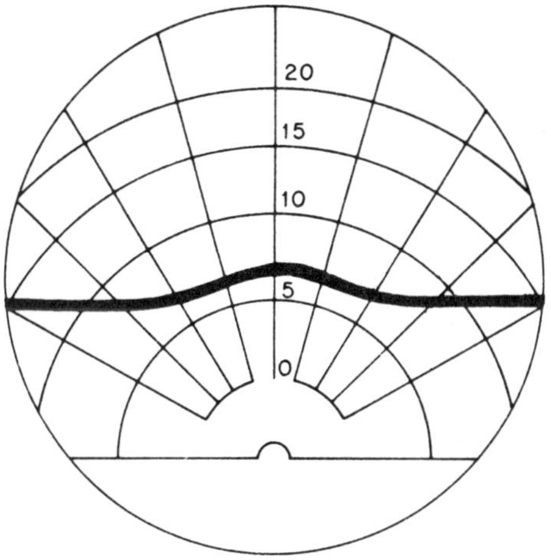

Fig. 13.3 Effect of slant range.

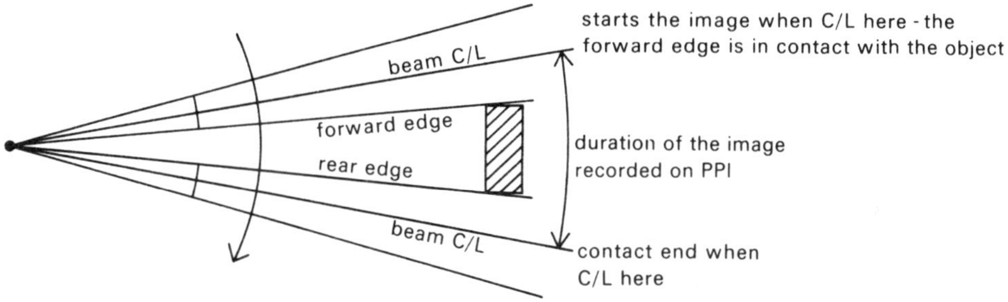

Fig. 13.4 Beamwidth and echo.

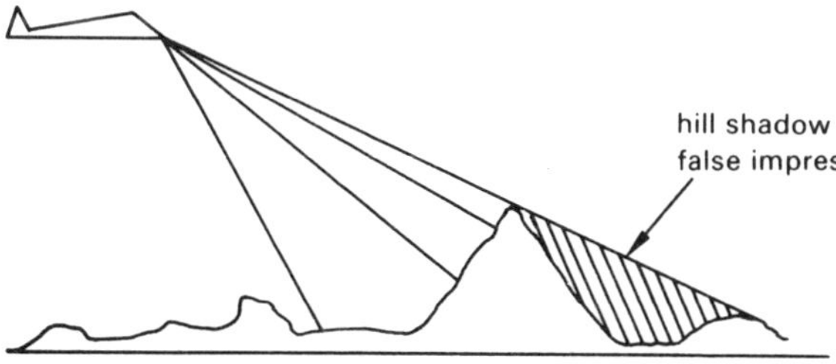

Fig. 13.5 Hill shadow.

are used to a route and seeing the same picture, it is worth bearing in mind that the picture can change substantially with seasons in the middle and higher latitudes. For example, rivers and lakes show up nicely in normal circumstances and also when iced up (ice has jagged edges which reflect), but will not show at all when covered with snow. The same is true for landmasses. Snow reduces the detection range of the equipment and the responses from the features are toned down, irregularities being removed by the snow deposits.

When using the equipment for weather detection, falling rain, wet hail or snow – in fact any liquid concentration – will show up. The three stages of a thunderstorm can be observed. The first appearance of the echoes indicates the beginning. The echoes grow in size and when the fuzzy edge is replaced by a well-defined edge the mature stage is reached. At this stage in monochrome a black hole will be seen when operating on contour function (see later). The dissipating stage is the process in reverse. The black hole gets smaller, ultimately disappearing, the cloud echo becoming ill-defined and weaker.

If a cloud shows up at a very long distance, say a hundred mile range, it indicates a presence of high liquid concentration – a cloud that must be watched.

Airborne equipment

The airborne equipment consists of the following units:

- transmitter/receiver
- an aerial scanning unit
- an indicator, and
- a control unit.

Transmitter/receiver

Transmission frequency varies with equipment within the range stated earlier. A typical AWR used by commercial airliners operates on a frequency of 9375 MHz (SHF or X-band). This gives a wavelength of 3.2 cm. The range covered is 320 nm. The mapping beam is 85° deep and scans a sector of 90° either side of the aircraft's centre line, ahead of the aircraft. With some equipment, a scan extending to 120° is available. The beamwidth of the conical pencil beams (weather beam) depends on the size of the scanner employed. An 18-inch scanner produces a beamwidth of 5° and a 24-inch scanner, 3.5°.

Aerial scanning unit

The scanner unit consists of a paraboloid dish with a centre dipole and the system is gyro stabilised in pitch and roll.

Indicator

A typical indicator is shown in Fig. 13.3. Bearings are marked on the face of the CRT with bearing lines at 15° intervals and the range markers are electronically produced. The intervals between range markers vary according to the selected range.

On a 150 nm range indicator:

- on 20 nm scale, range markers appear at 5 nm intervals
- on 50 nm scale, range markers appear at 10 nm intervals and
- on 150 nm scale, range markers appear at 50 nm intervals.

Similarly on a 120 nm equipment:

- on 20 nm scale, range markers appear at 5 nm intervals
- on 60 nm scale, range markers appear at 10 nm intervals and
- on 120 nm scale, range markers appear at 20 nm intervals.

If it is a colour CRT as in an EFIS multi-function display, typical ranges are 10, 25, 50, 100, 200 and 300 nm. Annunciator range markers are shown at 20%, 40%, 60% and 80% of the selected range. It is probable that the range and azimuth markers will be green on mapping and cyan (pale blue) on weather.

Control unit

A typical control unit is shown in Fig. 13.6 of a basic, stand-alone, monochrome weather radar.

The controls are usually:

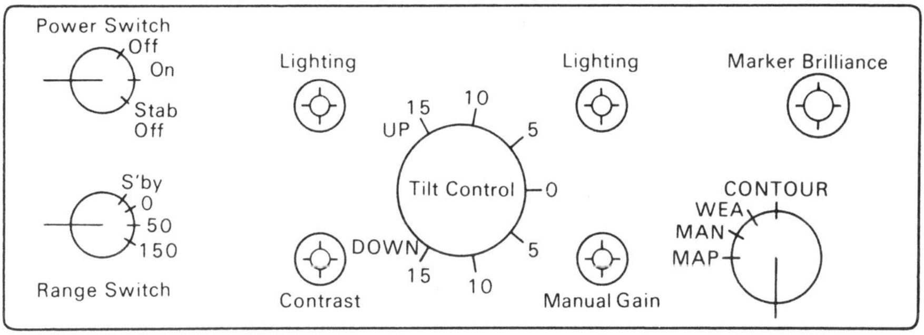

Fig. 13.6 Typical monochrome control unit.

Power switch. A three-position switch, it controls the power supply. In ON position the aerial is automatically stabilised, in STAB OFF position it is locked to the pitch and roll axes of the aircraft.

Time base range switch. This is a four-position rotary switch; in STBY position it maintains the equipment in readiness for instant use. Use this position when AWR is not required during short intervals.

Tilt control. Permits an aerial tilt from 0° to 15° UP or 15° DOWN. The aerial is normally tilted downward when ground mapping or it is raised up to estimate the heights of cloud base/top.

Function switch

MAP: In this position ground mapping is done by use of the cosecant beam. Manual gain control is used in conjunction with this position. This is because the signal strength can vary with altitude and type of terrain over which the flight is being made.

MAN: this is the next position after MAP and, again, is used for map painting. But the beam in use is the conical pencil beam and because of the concentration of energy inside a narrow beamwidth, the ranges obtained are greater than when operating on MAP position. And because ground mapping is being done, manual gain control is still operative in this position.

WEA: this is the normal position for observing weather. Manual gain control is inoperative; instead sensitive time control (STC) is brought into operation. It automatically reduces gain at short ranges so that with decreasing ranges the same target continues to produce the same contour separation, or all clouds at short ranges may be compared on equal terms.

CONTOUR: this position is used to examine the cloud structure for severity, and it shows up the turbulent areas on an iso-echo display (see below).

Marker brilliance. Varies the brightness of the range rings and azimuth lines which may be shown either as a 135° forward segment or as a 360° display around a central aircraft position.

The other controls (in Fig. 13.6) are self-explanatory.

On the latest models the controls are often positioned around the colour CRT and include the following pushon/pushoff switches:

TEST: this enables the test patterns of various bands of colour to be displayed for checking, after the system has been warmed up (after approximately 60 seconds).

HOLD: this allows the display to be frozen so that the storm movement can be assessed. See later paragraph on the procedure.

TGT ALERT: this operates in conjunction with Wx (or WEA) to notify the pilot of a thunderstorm return of contour strength. When TGT ALERT is selected and there is no relevant contouring weather ahead, the screen shows a yellow. T' in a red square in the top right corner. If a contouring target is detected ahead of the aircraft, typically within 60 nm and 160 nm and ±15° of dead ahead regardless of the range selected, the yellow letters TGT enclosed in a red rectangle flashing on/off once a second replace the 'T'.

AZ: this switch selects whether or not the five azimuth (or bearing) lines are displayed.

STAB: this is to stabilise the antenna in pitch and roll, and is usually kept on the ON position.

There will be a fault-monitoring circuit with a priority over any mode that has been selected. If there is a power interruption or the transmitter fails, the word FAULT flashes on the screen. The system is reset by momentarily selecting STBY and then the required mode.

Iso-echo display

Although the operator can see on the screen the complete cloud distribution ahead of him, without sufficient experience he would find it difficult to pick out dangerous clouds from among the less dangerous ones on a black/white display. The iso-echo type of display simplifies this distinction. It is known that strong turbulent clouds produce strong echoes compared with weak, inactive clouds, and this fact is utilised in the production of iso-echo display. All reflections above a predetermined echo level are cut off from reaching the screen. About the turbulent centre there is usually the remainder of the cloud whose activity is below the predetermined level. This will be seen on the screen. This signifies that a cloud on the screen with a hole in the centre is dangerous. The hole or the blacked out portion of the cloud is particularly dangerous, containing the highest concentration of large drops; the degree of the danger from the turbulence in the cloud depends mainly on the steepness of the contour (Fig. 13.7). On some screens, the hole may be drawn to the pilot's attention by being flashed cyclically.

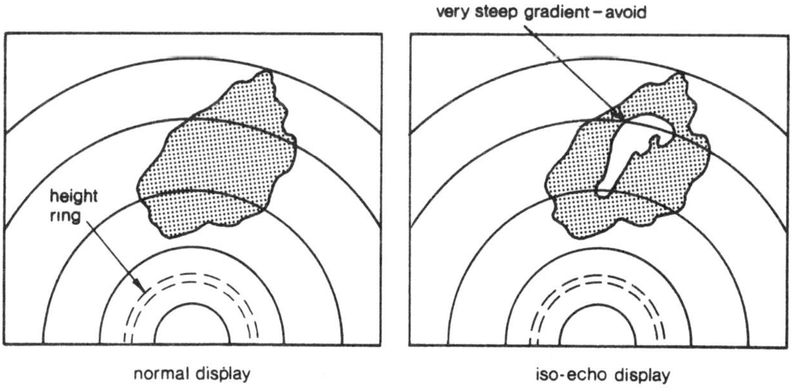

Fig. 13.7 Iso-echo.

Similarly on colour CRTs with a three or four colour presentation, increasing severity is shown by echo colours ranging from green to yellow to red and/or magenta for the areas of strongest echoes and heaviest rainfall.

The area of greatest turbulence occurs where the colour zones are closest together on the screen (i.e. the steepest contour gradient or paint).

Operation on ground

If the equipment is required to be switched on while on the ground take precautions, e.g., first check that there are no large hangar or other reflecting objects in the vicinity. The radiation can damage the health of personnel and the reflections can damage the equipment. And for the same reasons do not select mapping beam on the ground.

The usual pre-flight start-up check for a colour airborne weather radar is as follows:

(1) Check that the antenna is not pointed towards ground personnel, hangars or other large metal buildings, towards other aircraft or towards containers holding flammable material.

(2) Check that no fuelling or de-fuelling operations are being carried out.

(3) Check that the radar busbar circuit breaker is in.

(4) Select TEST mode.

(5) Select 25 nm range.

(6) Select GAIN control to MAX.

(7) Select TILT control to +5°.

(8) Leave the TARGET ALERT switch at OFF.

(9) Select AZ switch to ON.

(10) Select STAB switch to ON.

(11) Within approximately 20 seconds, the azimuth and range lines should appear together with the word TEST on the display. On a 25 nm range display, the range markers 5, 10, 15, 20 and 25 also appear.

(12) Within approximately 60 seconds, the test pattern will appear showing the colours in bands.

(13) Recheck that the antenna is *not* pointing towards personnel, hangars, large buildings, other aircraft or containers of flammable materials, THEN momentarily select WX (or WEA) mode, reselect TEST, select HOLD. The signs HOLD and TEST then alternate on the CRT, superimposed on the frozen test pattern.

(14) Select Wx (or WEA) when that will appear on the screen.

(15) Select 10 nm range and note the new range rings and their annunciators.

(16) Adjust the tilt control over the whole range noting ground returns

at the low settings and any thundery weather returns at the higher settings.

(17) Select STBY mode until airborne.

Operating on weather

Select maximum range to detect presence of clouds in good time. When the clouds close up to around 50–60 nm, select lower scale and operate on CONTOUR function. Ensure from the beginning that no ground returns are being received – the aerial tilt may be adjusted to exclude them. Any storms detected at very long ranges are potentially dangerous. Take avoiding action when the range gets below 20 nm. If possible avoid all response areas; if this cannot be done, then at least avoid responses with black holes or red/magenta areas.

Apart from the points already made under iso-echo display, there are other characteristics of the returns of which pilots should be aware. Hail associated with thunderstorms may well produce 'fingers' or stubs out from the main echo up to 5 nm long. Other shapes can be in the form of hooks, scalloped edges and U-shapes which even if there is no hail can usually be relied upon as good indicators of areas of severe turbulence. The extreme case of a thunderstorm is a tornado associated with cumulonimbus-mammatus. In the same way as the actual tornado cloud has a writhing funnel cloud beneath it, the tornado cloud echo often shows a narrow, finger-like extension from its main outline, which quickly curls into a hook and then closes on itself. Fortunately, tornadoes are infrequent in Europe and indeed most of the world but they are common in the USA.

Storm movement

From basic meteorological principles, cumulonimbus clouds and thunderstorms move in the direction of the 700 mb flow (approximately the 10 000 ft winds). Self-propagating storms usually generate on the forward side of the storm. When fitted, the HOLD facility enables the storm movement to be assessed by the flight crew. The procedure when, say, a storm is detected on the 100 mile range, is to select the push/push HOLD switch while maintaining constant heading. HOLD and Wx then appear alternately on the screen. After a couple of minutes, select the push/push HOLD switch again. This brings back the current display and the storm echo moves from its previously held position to the current position so showing its movement relative to the aircraft.

Operating on mapping

Cosecant beam is used for short range mapping (up to 70 nm). To obtain optimum ranges use pencil beam for reasons given earlier. Give the aerial a large downward tilt initially and then reduce the tilt angle very slowly, keeping a watch on the changing coverage. When no further forward coverage is observed, you have the right tilt for maximum ranges.

Operating in conjunction with RNAV systems

The capacity of the modern EFIS computers can also be used in conjunction with AWR displays. The output from the navigation management system can be superimposed on the colour display as the track between waypoints, as a pale blue line on the multi-function display.

Calculating approximate height of the cloud

This is done by tilting the beam upward or downward as necessary until the cloud just disappears from the screen. At this time, the base of the cone is directly on top of the cloud. This gives the angular measure of the cloud height above or below the aircraft's level. We also know the range of the cloud. Thus, we have sufficient information to apply the 1 in 60 rule to calculate the height above/below the aircraft level (Fig. 13.8).

For example, it is required to estimate the height of the top of an active cloud, 30 nm range.

The first step is to tilt the aerial (Fig. 13.8) until the cloud top just disappears. Say, at this time the tilt angle is $3\frac{1}{2}°$. This $3\frac{1}{2}°$ defines the centre line of the beam. Therefore, assuming 5° beam width, the base of the beam is 1° above the aircraft's horizontal plane. Now the data are put on computer as follows:

$$\text{Angle above aircraft} = \frac{\text{Dist off (rel ht)} \times 60}{\text{Dist}}$$

$$1° = \frac{\text{ht} \times 60}{30}$$

i.e.

$$\text{ht} = \frac{1 \times 30}{60}$$

$$= \tfrac{1}{2}\,\text{nm or 3040 ft}$$

This is the height above the aircraft height. Alternatively, the makers provide pre-computed graphs from which the height is read off against tilt angle.

A rough estimate of the height may be made mentally from the formula: relative height = 100 × range in nm × elevation in degrees. In our illustration it is 100 × 30 × 1 = 3000 ft.

Height ring

With conical radiation there is an overspill of radiation vertically downwards (Fig. 13.8). This is reflected back to the aircraft and the echo appears like a ring. This ring indicates the aircraft's height and therefore it is called the

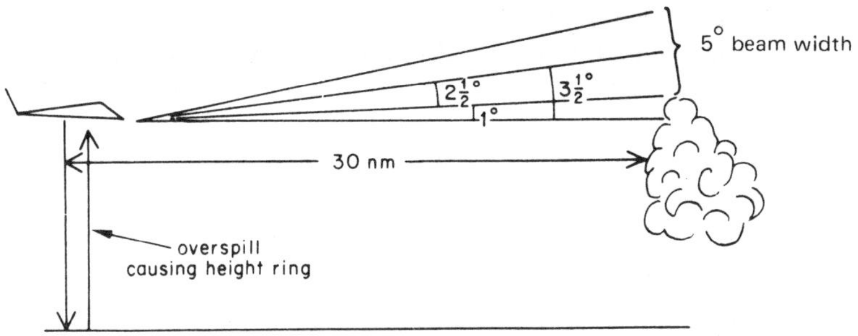

Fig. 13.8 Finding cloud height using 1 in 60 rule.

height ring. The presence of the height ring indicates that the equipment is serviceable. Its absence normally means that the receiver requires retuning. It may be possible to retune the receiver in flight – follow the manufacturer's drill and if the height ring is still not visible, the equipment may be considered unserviceable. It should, however, be remembered that there are few reflections when flying over calm water or fine sand particularly when flying above 30 000 ft. Also, as 4.93 nm equals 30 000 ft, if you are flying at that height, the height ring may merge with the 5 nm range marker. To ensure its presence, turn the marker brilliance down. Finally, it is also possible to see a second height ring, at twice the aircraft height. If it is present, it will be faint because it is due to double reflections.

Other uses of AWR

The equipment may be used to ensure a safe terrain clearance over high ground. By tilting the aerial slightly downward, keep a watch as the high ground gets closer. If it does not disappear from the picture at a reasonable range, a climb must be initiated immediately.

Test questions

(1) How does the use of the iso-echo function on an airborne weather radar enable a pilot to detect areas of severe turbulence?

(2) What are the main factors which determine the most suitable wavelength to be used in airborne weather radar? Say why.

(3) Using AWR, the conical beam should be used for map painting in preference to the fan-shaped beam when:
- (a) there are thunderstorms in the vicinity
- (b) maximum range is required to be scanned
- (c) approaching a coastline.

(4) AWR equipment operation on the ground is:
- (a) totally prohibited
- (b) unrestrictedly permitted in aerodrome maintenance areas
- (c) only permitted with certain precautions, to safeguard the health of personnel and to protect equipment.

(5) AWR operates in the SHF band because:
 (a) large water droplets, hailstones, etc., give good reflections at a 3 cm wavelength, while small droplets of fog, mist, or stratus do not
 (b) it enables equipment to be powerful but light enough to be installed in the aircraft nosecone
 (c) it enables a narrow width beam to be used to obtain the required degree of definition.

(6) The main purpose of selecting HOLD on an AWR is:
 (a) the storm movement can be assessed
 (b) the storm potential development can be assessed
 (c) the storm's potential rate of decay can be assessed.

(7) Hail associated with thunderstorms will often be shown on an AWR as:
 (a) an extremely straight edge to the echo
 (b) a finger extending out from the echo
 (c) a distinct concave edge to the echo.

(8) On an AWR the heaviest precipitation is shown by:
 (a) green echoes (b) yellow echoes (c) red echoes.

(9) The principle of weather radar is that turbulence is always greatest where:
 (a) there is the strongest contour gradient
 (b) hail and the largest size raindrops are detected
 (c) lightning is most frequent.

(10) To measure the movement of a thunderstorm when using an AWR, select:
 (a) the MOTION switch (b) the HOLD switch (c) the TRACK switch.

Chapter 14
RADIO ALTIMETERS

FM altimeters

Radio altimeters are designed to give indication of actual height above the surface. Those operating on the principle of FM are mostly used for low level flying where a high degree of accuracy is required. They are otherwise unsuitable for level flight since they indicate the contour of the ground below instead of a constant level. When installed, however, they may be used to check the accuracy of a pressure altimeter if the elevation of the ground below is known. In the past much use was made of them in flying 'pressure pattern'. In the latest development, they can be used to feed 'height' information to an automatic landing system.

Principle

The principle of FM is utilised to measure time (and thereby, height) taken by a radio wave to travel to the surface directly beneath, and to return. They operate in the band 4200 to 4400 MHz.

The aircraft transmits radio signals vertically downward. The transmission frequency, however, is not constant but it is varied progressively and *at a known rate* from its start frequency to 'start + 60' MHz, and then back to the start frequency. This constitutes one complete cycle of modulation.

Some of these transmitted signals will be reflected and return to the aircraft. The frequency of these returning signals, however, will be different from the frequency actually being transmitted at that instant. Further, the higher the aircraft, the greater the difference between the two. Since the rate of change of frequency is fixed this difference must be proportional to the time taken, that is, the height of the aircraft, as the speed of the radio propagation is known.

$$\text{Difference in frequency} = \text{rate of change of frequency} \times \text{time taken}$$

This frequency difference is measured and the time obtained is registered on the indicator as height.

Equipment

Transmitter/receiver. These generally work in conjunction with separate transmitter/receiver aerials.

Indicator. Several different types of indicators are available. One of them, shown in Fig. 14.1, has a pointer on the face of the dial calibrated in hundreds of feet. Thousands of feet indication is given on a counter in the window at the 6 o'clock position. The instrument shown reads 1360 ft.

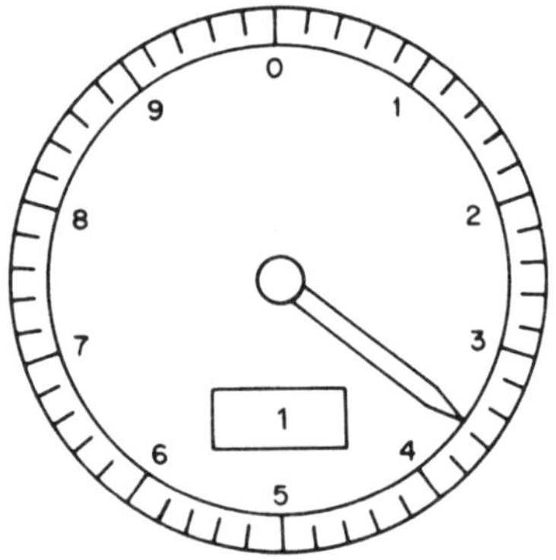

Fig. 14.1 Simple radio altimeter dial.

The indication is not limited to one single indicator in one position. As many repeater indicators as necessary may be installed to cater for various crew positions.

Although most indicators have a pointer and dial calibrated on a linear scale, many displays have a special arrangement for the scale at the lowest altitudes. For example, from 0 to 500 ft may be shown on an expanded scale Fig. 14.2. Other displays have not only an expanded scale to 500 ft, but also a

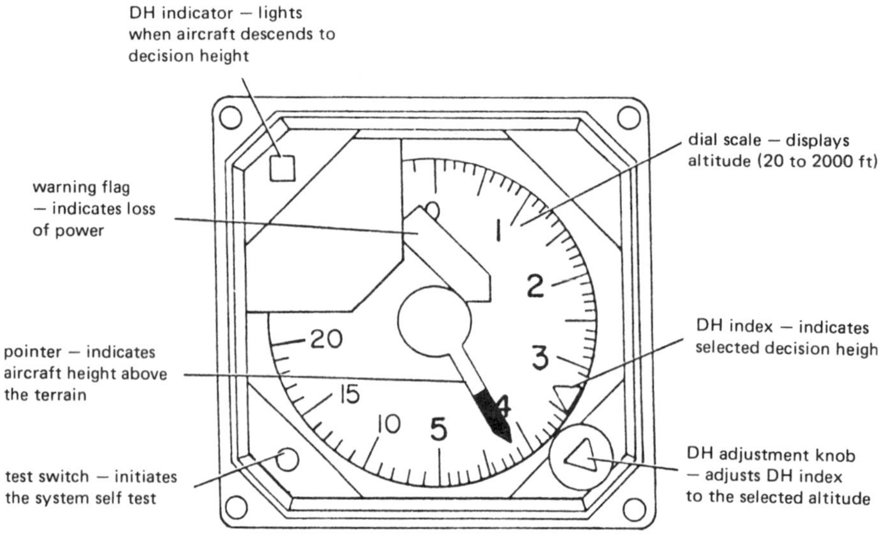

Fig. 14.2 Radio altimeter/DH indicator.

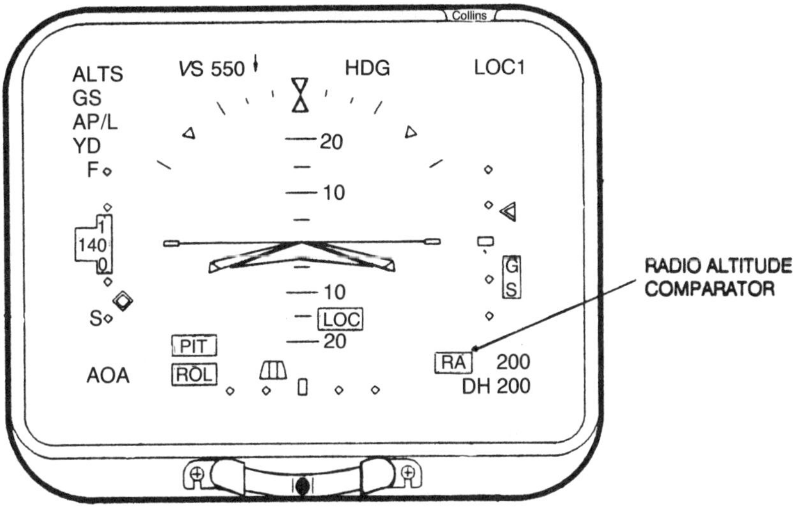

Fig. 14.3

logarithmic scale from 500 to 2500 ft. Digital displays are available, while with an EFIS (*volume 3*), a comparative display of radio altitude and decision height are shown on the EADI (Fig. 14.3).

Decision height indicator

Many indicators are equipped with a DH selection/control knob (see Fig. 14.2). The DH control knob may incorporate a press-to-test facility. Either pre-flight or in-flight, the DH knob is used to position the DH bug at the required DH setting. When the aircraft descends below the set decision height, the DH lamp comes on, and in some systems an audio tone is also produced.

Uses

Mainly when flying at low level. Most instruments cater for height up to 2000 or 2500 ft and are compatible with GPWS (Chapter 15).

Accuracy

Fixed error. This error arises in the method of transforming frequency difference into height and feeding the current to the indicator. The indicator pointer moves in steps of 5 ft, which means an error up to $2\frac{1}{2}$ ft may be present any time. Overall accuracy is of the order of 5 ft $\pm$ 3% of the indicated height.

In some installations, the radio altimeter may indicate less than zero altitude when the aircraft is parked on the ramp. This is because the altimeter reads zero ft on the point of touchdown and where the antenna is nearer the ground when parked.

Pulse modulation altimeters

These altimeters operate in the frequency band of 1600 MHz. Originally height was indicated by a blip on a circular time base, and was read off in terms of its distance from the start of the time base which also appears as a blip. On current equipments the flight-deck display is of the pointer and dial type already described for the frequency modulation altimeter (Fig. 14.2).

Principle

The principle utilised is the echo. The time taken by a radio pulse to travel out and back is the direct measure of the distance (height in this case) on the assumption that the speed of the radio wave is constant. This assumption is not an unfair one, since the speed of a radio wave is known to a considerable degree of accuracy. The timing, of course, must be done electronically by the process of leading edge tracking.

The master circuit instructs the transmitter to transmit, and at the same instant, starts the timing process. The transmission takes place vertically downward in the form of a series of pulses in a wide conical beam from the antenna. Signals reflected from the ground underneath will be picked up by the receiver aerial, provided that the aircraft is not banking excessively (usually of the order of 40°).

Limitations

The maximum range of modern equipments (Fig. 14.2) depends upon the surface reflectivity. The best surface reflector is water, whilst the worst is dry, fine, loose soil. Hard-packed earth gives better reflection than snow. Nevertheless the manufacturers claim that the height is measured from the top of the snow cover. Such indicators may have a loss of lock due to lack of reflectivity at altitude (say 2000 to 2500 ft), either due to a steep bank or to the nature of the terrain beneath the aircraft. Under the impetus of developments required for military purposes, there has been a steady improvement in the accuracy of radio/radar altimeters. Most have a capability of reading down to 20 ft and an accuracy of ±2 ft or 2% up to 500 ft.

Advantages of radio/radar altimeters

The radio/radar altimeter has the following advantages:

(1) It gives actual height above the ground – a very useful piece of information when flying low level.
(2) It provides a cross check capability with the pressure altimeter for terrain clearance purposes.
(3) Models working in the SHF frequency band of 4200–4400 MHz produce

very high accuracy (2 ft or 2% from 500 ft to touchdown). Developed for operation with categories 2 and 3 ILS, they provide a tie-in with flight director/autopilot to initiate initial flare during approach and landing.

(4) A warning signal at DH is usually given. As will be described in chapter 15, some models of GPWS provide a warning MINIMUMS when the aircraft descends through the DH setting that has been made by the pilot.

Test questions

(1) Radio altimeters operate in the:
(a) VHF band (b) SHF band (c) VLF band.

(2) Radio altimeters operate up to:
(a) 2500 ft (b) 1500 ft (c) 500 ft.

(3) The mean frequency on which radio altimeters operate is:
(a) 4200 MHz (b) 4300 MHz (c) 4400 MHz.

(4) A DH setting knob and 'DH set' indicator:
(a) can only be integrated in a radio altimeter if it is part of an EFIS
(b) is usually incorporated in modern radio altimeters
(c) has to be a separate installation to the radio altimeter itself.

(5) The formula used to determine the height displayed on the radio altimeter is that the difference in frequency equals the rate of change of frequency:
(a) times half the time taken
(b) times the time taken
(c) times twice the time taken.

Chapter 15
GROUND PROXIMITY WARNING SYSTEM (GPWS)

This highly desirable piece of flight-deck equipment came into general use in the 1970s with the development of microprocessors and sophisticated voice synthesisers. It has been estimated from the GPWS statistics that possibly as many as 60 accidents have been averted in the first decade of its installation. Mandatory carriage of ground proximity warning systems has been introduced in many countries. In the UK the CAA has published *CAP 516 Guidance material on GPWS*.

The purpose of the system is to give visual and audible warning signals to the pilot when the aircraft is entering a potentially dangerous ground proximity situation. Such situations as inadvertent sinking after take-off, inadequate terrain clearance, excessive rate of terrain closure, and dropping below the correct glidepath when using ILS, would all activate the GPWS to alert the pilot to the potential danger. It is a system to enhance safety and not a foolproof means of preventing collision with the ground.

Terminology

Alert: a caution generated by the GPWS equipment.

Warning: a command generated by the GPWS equipment. It may be variously described as genuine, nuisance or false.

Genuine warning: one generated by GPWS in accordance with its technical specifications.

Nuisance warning: one generated by the GPWS in accordance with its technical specifications but the pilot is actually flying an accepted safe procedure.

False warning: one generated by the GPWS not in accordance with its technical specifications due to a fault or failure in the system.

The term 'unwanted' may be used to describe both nuisance and false warnings. The limits of protection provided by the aircraft's GPWS and other relevant details are included in the Operations Manual, Training Manual and checklists.

Equipment

Input. The basic equipment comprises a small digital computer or central processing unit (CPU) which accepts inputs from:

- the radio altimeter with failure signal
- vertical speed sensor or a barometric altitude rate computer with failure signal
- the ILS glide path receiver
- switch activated when the landing gear is down or is selected down
- switch activated by any crew selection uniquely associated with final approach to landing (usually flaps selection to, or are in, the landing position).

Output. If, after assessment, a potential danger of colliding with terrain is found to exist, the CPU will put out warning signals to the pilot in both visual and audible forms. The CPU will also put out indications of computer failure and any failures which may occur in the five input signals, to a monitor indicator. A block schematic diagram of how the system works is shown in Fig. 15.1.

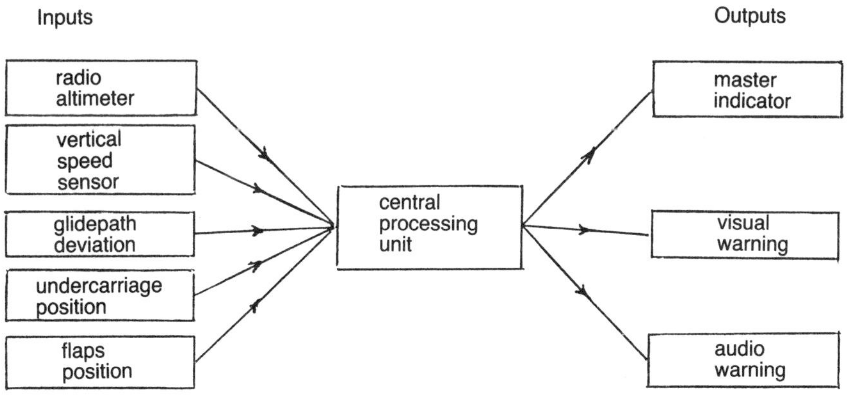

Fig. 15.1 Elements of Ground proximity warning system

Integrity testing. GPWS should be serviceable at take-off and the Operations Manual should detail the pre-flight checks verifying that the system is functioning correctly. Minimum equipment or allowable deficiencies lists should indicate when flights can be made with an unserviceable GPWS although legislation provides that an aircraft may fly, or continue to fly, with an unserviceable GPWS until it first lands at an aerodrome where it is practicable for the equipment to be repaired or replaced. GPWSs, like most of the other equipments described in this volume, have a fully integrated self-test function (built-in test equipment – BITE) for checking out the signal path from all of the inputs in the pre-flight test. When the BITE is selected, if the GPWS is satisfactory the normal indication to the pilot is that both the visual and audible warnings are activated simultaneously. The BITE is normally inhibited from being operative whilst in flight.

Modes of operation

GPWSs monitor six basic modes of the aircraft's operation and put out warnings as shown in Table 15.1 if a hazardous situation is arising. The equipment design is such that it automatically selects with no action from the flight-deck crew, Mode 3 for take-off or go-around below 500 ft and Mode 4 for landing.

Monitoring facilities

The system is continuously monitored in flight to confirm dynamic performance. Any failure occurring within the system is automatically indicated on the flight-deck.

Actions

The formal instructions to flight crews on their actions in response to GPWS warnings are given in the Operations Manual.

Alert: When the GPWS produces an alert which is regarded as a caution, the pilot must respond immediately by correcting the flight path or the aircraft's configuration so that the alert ceases. Alerts are associated with modes as:

Mode 1	Sink rate
Mode 2	Terrain
Mode 3	Don't sink
Mode 4A	Too low – gear
Mode 4B	Too low – flaps
Mode 5	Glideslope

Warning: When the GPWS generates a warning, the pilot should immediately respond by levelling the wings and initiating a maximum gradient climb, maintaining the climb until the minimum safe altitude is reached. Typically such warnings would be:

Pull up	Mode 1 or
Too low – terrain	Mode 4A or Mode 4B

Table 15.1

Mode	Flight hazard	Warning envelope	Alert	Warning
1	Excessive rate of descent with respect to terrain	50 ft–2450 ft	SINK RATE	WHOOP WHOOP – PULL UP + flashing lamp
2A	Excessive rate of terrain closure (aircraft not in a landing configuration)	50 ft–1800 ft	TERRAIN TERRAIN	As mode 1
2B	Excessive rate of terrain closure (aircraft in a landing configuration)	220 ft–790 ft	TERRAIN TERRAIN	As mode 1
3	Negative rate of climb after take-off or go-around	50 ft–700 ft	DON'T SINK	—
4A	Unsafe terrain clearance (aircraft not in a landing configuration – Gear up)	50 ft–500 ft	TOO LOW – GEAR	As mode 1
4B	Unsafe terrain clearance (aircraft not in a landing configuration – Flaps up)	50 ft–500 ft	TOO LOW – FLAPS	'TOO LOW – TERRAIN'
5	Excessive downward departure from ILS glideslope	100 ft–500 ft	GLIDESLOPE	—
6	Descent below minimums	—	MINIMUMS	—

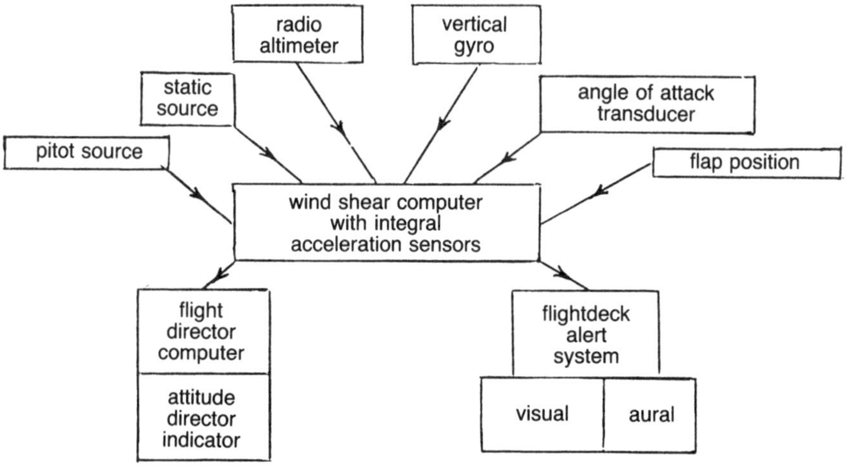

Fig. 15.2 Block diagram: windshear detection and recovery guidance system.

There may be an exceptional circumstance such as when on a curved approach or on a missed approach procedure involving a turn to avoid terrain when the requirement to level the wings would be inappropriate.

General

(1) All alerts and warnings should be responded to by pilots immediately. However in the case of a warning in the following circumstances only, a pilot can limit his response to that of an alert when:
  (a) the aircraft is operating by day in meteorological conditions which will enable the aircraft to remain 1 nm horizontally and 1000 ft vertically away from cloud and in a flight visibility of at least 8 km; and
  (b) it is immediately obvious to the aircraft commander that the aircraft is not in a dangerous situation with regard to terrain, aircraft configuration or the present manoeuvre of the aircraft.

(2) It must be remembered that GPWS is not a forward-looking equipment. Any warning that the aircraft is flying towards high ground will be minimal and dependent upon the steepness of the rising ground. Hence a sheer cliff, escarpment or mountainside will not generate any warning at all.

(3) Unwanted warnings still occur under normal and safe operating conditions such as when aircraft are carrying out standard terminal approach procedures or radar vectoring approaches at aerodromes surrounded by hilly terrain.

(4) The Operations Manual specifies conditions when the flight crew may inhibit the GPWS. Typically these would cover when gear or flap position inputs are known to be non-standard or if the aircraft is being flown

without reference to the glideslope such as when it breaks off an approach to an ILS runway so as to land on another runway.

(5) GPWS warnings are normally reportable occurrences. GPWS alerts are only reportable occurrences when a hazard arises or when there is undue repetition.

Windshear detection

The serious hazards associated with low-level windshear and microbursts (*see Ground Studies for Pilots volume 4*) have been extensively explored in recent years. Equipment has been developed to provide a specific windshear detection and recovery guidance system (WDGS) operating through the flight director system.

Typically the windshear is detected from a comparison between the rate of change of groundspeed (measured by accelerometer) and the rate of change of airspeed (measured by pitot and static sensors). A shear adversely affecting performance illuminates a red warning light on the panel and gives an aural warning WINDSHEAR WINDSHEAR. The system has a WINDSHEAR FAIL annunciator which illuminates in the event of a power loss.

The recovery guidance for an adverse effect on aircraft performance is based on angle of attack, modified by radio altitude and pitch attitude and is displayed via the flight director, through the pilot's attitude director indicator (ADI) or the EFIS primary flight display (PFD).

Chapter 16
DOPPLER

There are a number of stories about Doppler saying that after he noticed the change in pitch of a passing train's whistle he was stimulated by the challenge and produced his theory of frequency shift. The fact of the matter is, Dr C. J. Doppler was a scientist and a Professor of Experimental Physics in Vienna and he was engaged on a research project to establish the relationship between waveform and motion. His primary interest was to calculate movements of the stars and it was to this end that the theory was immediately put into practice. In 1842 he published his work in a paper entitled *Uber das farbige Licht der Doppelsterne* in which he explained the relationship between the frequency of a waveform and the relative motion of an object.

The paper establishes that whenever there is a relative motion between a transmitter and a receiver, a frequency shift occurs. This frequency shift is proportional to the relative movement. It is now variously known as:

Doppler shift, or
Doppler frequency, or
Doppler effect

and is abbreviated f_d. If the distance is closing, the received frequency is greater than the transmitted frequency and it is described as a positive Doppler shift. Similarly, if the objects are moving apart, the received frequency will be smaller and it is a negative Doppler shift.

Why does frequency shift take place?

Consider a radio signal, transmitted from a stationary source at a carrier frequency of f Hz. The receiving object, R will receive f waveforms each second, each say λ metres apart. Further, they are approaching R at the speed of electromagnetic waves, say c metres per second.

Thus, R receives a frequency which is equal to c/λ which in fact is the transmission frequency. Now suppose that the object R is moving towards the transmitter at a speed of V metres/second. Although the waves still travel at the same speed, because of the object's motion towards the source of transmission, each wave will arrive at the receiver at a progressively shorter time interval than its predecessor, the first wave taking the longest time and the last one the shortest time to reach the receiver. This progressive closing up of the waves is sensed by the receiver as a reduction in the wavelength.

Now, the relationship between speed of electromagnetic waves, frequency and wavelength is expressed in the formula $c = f \times \lambda$ (see chapter 1) where c is constant. Therefore, if λ decreases (as it does when the objects are closing) the value of f must increase to keep the value of c constant. This frequency shift is entirely due to the relative motion between the transmitter and the receiver.

The transmission frequency is c/λ.

$$\text{The received frequency} = \frac{c + V}{\lambda}$$

$$= \frac{c}{\lambda} + \frac{V}{\lambda}$$

$$= f + \frac{V}{\lambda}$$

f being the original transmission frequency, V/λ is the component due to the relative motion between the transmitter and the receiver and is the Doppler shift, which is proportional to the ground speed of R. We express this in the formula

$$f_R = f + \frac{V}{\lambda} \qquad (1)$$

where f_R is the received frequency. Note that because

$$\lambda = \frac{c}{f} \quad \text{and} \quad \frac{1}{\lambda} = \frac{f}{c}$$

this formula may be rewritten as $f_R = f + \frac{Vf}{c}$.

Adapting the formula for aircraft operation

Initially, the above formula is modified in two ways.

(1) In the case of an aircraft, the signals are transmitted from a moving transmitter and are reflected back to a moving receiver. Hence, the relative velocity is two-fold or $2V$ and the formula is modified to read

$$f_R = f + \frac{2V}{\lambda} \qquad (2)$$

(2) The above formula is applicable only when the transmission is directly ahead and the reflection occurs from a reflecting object, also directly ahead. Although such transmission and reception would in theory give a maximum frequency shift in, practice a pilot will not be operating with Doppler with a solid reflecting object directly in front of him. Therefore, Doppler equipment must beam its signals towards the earth's surface for reflections. The formula now becomes

$$f_R = f + \frac{2V\cos\theta}{\lambda} \tag{3}$$

where θ is the angle through which the beam is depressed.

Depression angle

As pointed out above, ideally the signals should be beamed straight ahead for maximum frequency shift. This is, however, impossible. Therefore, the signals must be beamed downward towards the surface at an angle from the horizontal.

The angle of depression may be kept small, for example, to achieve a good measure of the frequency shift. However a signal striking the surface at a shallow angle loses quite a lot of energy by way of scatter *away* from the aircraft (see Fig. 16.1). If the depression angle is made too large on the other

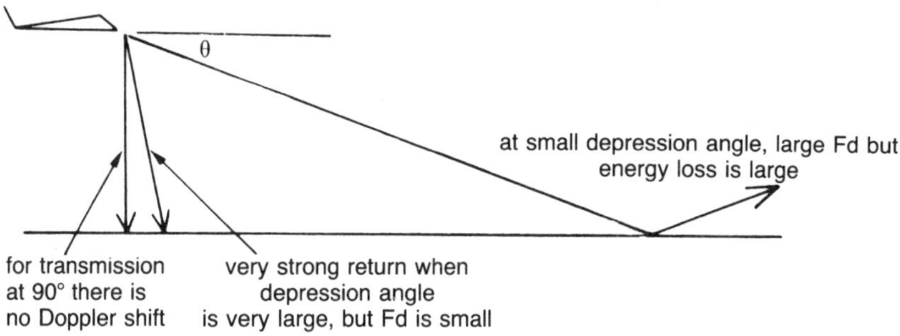

Fig. 16.1 Effect of different angles of depression.

hand, strong reflections will be obtained but the value of f_d, that is, $2Vf\cos\theta/c$ becomes too small for accurate measurement. The value of cosine decreases as the angle is increased and if the transmission was vertically downward, the cosine of 90° is 0 and there will be no frequency shift.

Therefore, the depression angle of Doppler equipment is a compromise between strength and frequency shift. Or, it must be an angle which would ideally give the best of both. Most Dopplers operate between a depression angle of 60° and 70°.

Principle of ground speed measurement

Electromagnetic waves are transmitted from an aircraft at a given depression angle either in the form of a series of pulses (pulse radar) or continuous wave (CW radar). On striking the surface these will be reflected. Some of these reflected signals will be received in the aircraft receiver but the frequency of these signals will be different from the original transmission frequency. It will be higher if the signals are reflected from forward of the aircraft, and lower if reflected from the rear of the aircraft. This change in the frequency is due to the Doppler shift and because it occurs due to the relative movement it expresses the aircraft's GS. This is continuously indicated on the indicator. The GS displayed is 'spot' GS, that is, correct only at that instant.

Doppler aerials

Single beam systems. These represent the earliest thoughts on aerial deployment with Doppler equipment and are no longer in use. A beam is transmitted to the fore of the aircraft to the surface. When the reflected signals are received and f_d is available, the aerial is rotated until the value of f_d is maximum. The angle through which the beam is thus displaced is the measure of drift and the f_d is the direct measure of the aircraft's ground speed.

Twin beam systems. These may be fixed or rotatable beams.

(1) *Fixed aerials.* One beam is directed forward, the other abeam of the aircraft. The f_d produced by the two beams thus disposed represents the aircraft's speed along its heading and at right angles to it. The two values are compounded to give drift and GS.
(2) *Rotating aerials.* The two beams at a fixed angular distance are rotated about the fore-and-aft axis of the aircraft until the f_d produced by both the beams is of equal value. Then the bisector of the angle between the two beams is the aircraft's track and the ground speed is calculated from the extension of formula 3

$$f_d = \frac{2Vf}{c} \cos\theta \cos\phi \qquad (4)$$

where ϕ is half angle between azimuth beams, i.e. skew angle, a known fixed value.

Three-beam systems. Where three beams are employed, two beams are directed to right and left forward and the third beam almost vertically downwards, as shown in Fig. 16.2. Printed antennae for transmission and reception are mounted under fibreglass radomes on the underside of the fuselage. A single transmitter is switched sequentially into each of the three beams in the Janus configuration, the sequence being repeated approximately ten times a second.

Modern three-beam systems have the advantages of:

- reduction of number of components
- economies of weight
- economies of cost
- greater reliability
- improved presentation of information on the flight-deck.

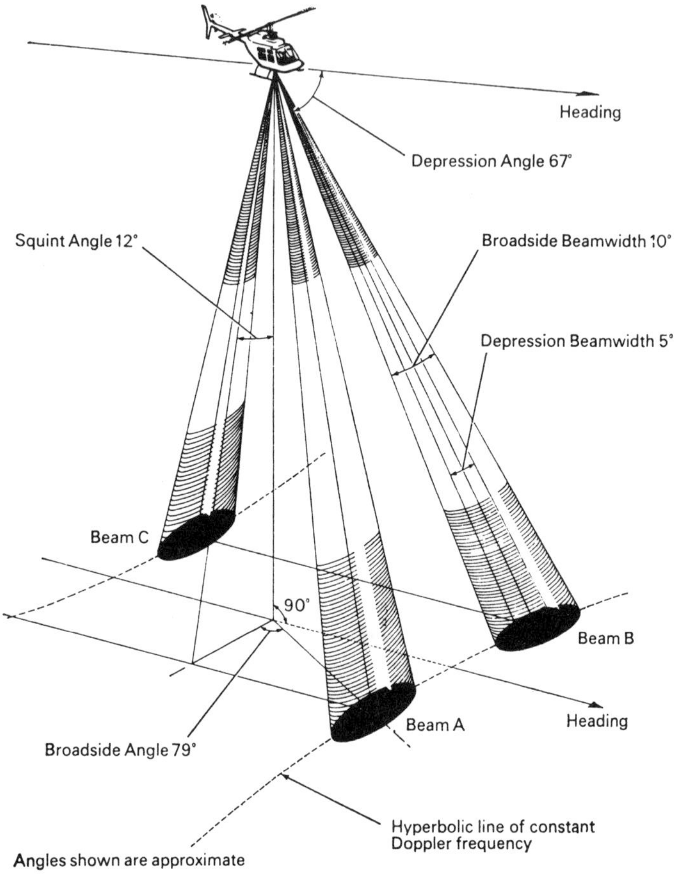

Fig. 16.2 Three-beam system.

Separate versions apply to rotary-wing aircraft (typically for speeds up to 350 kt, altitudes up to 25 000 ft) and fixed-wing aircraft (typically for speeds up to 1000 kt, altitudes up to 60 000 ft), the system enabling the calculation not only of the horizontal velocity components but of the vertical velocity components as well.

Four-beam systems

(1) *Fixed aerials.* Four beams are transmitted from two transmitters. One transmitter transmits beams 1 and 3 (see Fig. 16.3) simultaneously. Beam 1 is directed to forward port and beam 3 to rearward starboard. Half a second later the second transmitter transmits beams 2 and 4 to forward starboard and rearward port. Beams 1 and 3 produce a component of ground speed in direction forward port and beams 2 and 4

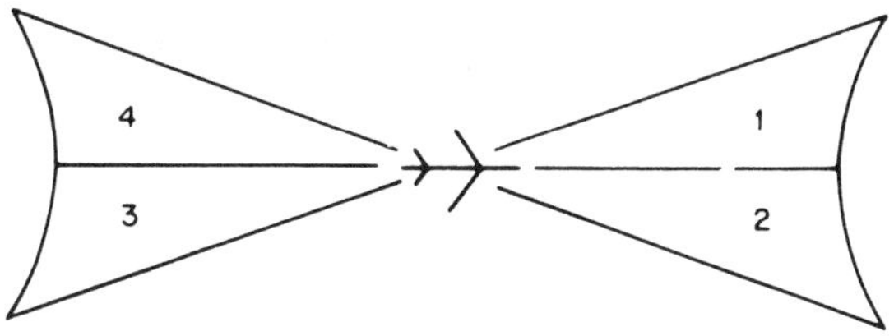

Fig. 16.3 Switching pattern.

in direction forward starboard. Since the angle between the beams is known, drift and ground speed can be computed.

(2) *Rotating aerials*. This is perhaps the most popular system. The transmission pattern is similar to the four-beam fixed aerial system. One aerial sends a beam forward and to the right while the other aerial of the pair sends the beam rearward and to the left. A few seconds later the other pair of aerials sends forward left and rearward right. The returning signals from each pair are mixed together to produce a Doppler beat frequency.

When the centre line of the aerial system is along the fore-and-aft axis of the aircraft, and there is no drift, the speed of the aircraft towards both forward reflections is the same. The Doppler frequencies produced by each pair will be identical.

If it has drift, the beat frequency extracted from 1 and 3 aerials and 2 and 4 aerials are explained with reference to Figs. 16.4 and 16.5.

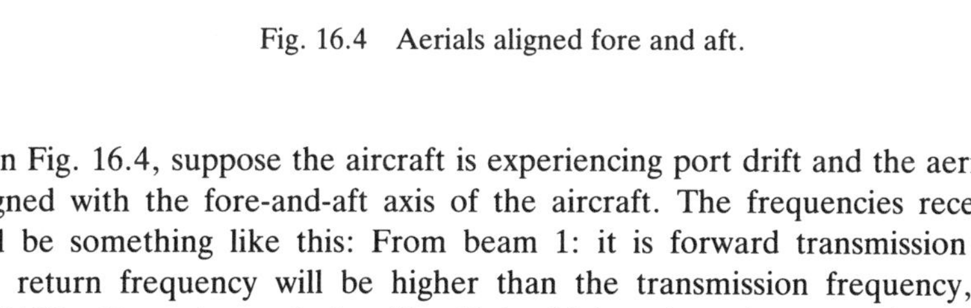

Fig. 16.4 Aerials aligned fore and aft.

In Fig. 16.4, suppose the aircraft is experiencing port drift and the aerial is aligned with the fore-and-aft axis of the aircraft. The frequencies received will be something like this: From beam 1: it is forward transmission and the return frequency will be higher than the transmission frequency, say +10 kHz. From beam 2, it will still be higher than the transmission frequency, but not as high as the return from beam 1, since the aircraft is drifting away from beam 2 to beam 1. Say, the frequency received is +8 kHz. Beam 3 will produce −10 kHz and beam 4−8 kHz. Beat note from beams 1 and 3 = +10 − (−10) = +20. Beat note from 2 and 4 = +16. This difference between the two is entirely due to the direction of the aircraft aerial and the aircraft's track not being the same. As soon as such a difference arises between the two beat notes, a signal is raised which actuates a motor. The motor turns in the direction of the track, turning the aerials with it and will continue to turn until the difference between the two pairs of aerials is reduced to zero. At this time the aerial is aligned with the aircraft's track (Fig. 16.5) and the motor switches off.

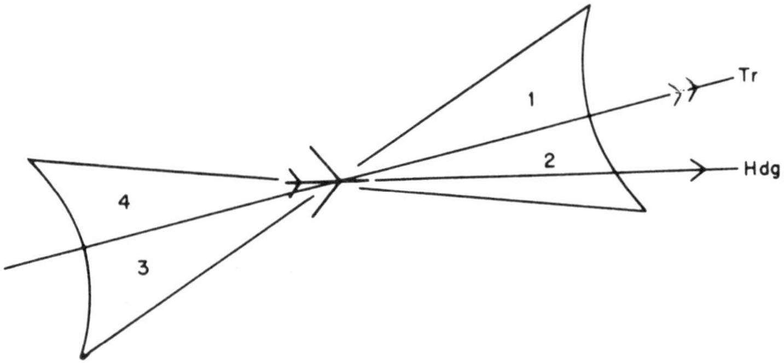

Fig. 16.5 Aerials aligned with track.

The angular move of the aerial from the fore-and-aft axis of the aircraft is the drift which is indicated on a suitable indicator. (It will be noticed in Fig. 16.5 that the aerial is aligned with the track and the Doppler spectrum is symmetrical about the track axis.) This is a condition for the ground speed measurement to be correct.

Thus, drift is measured directly and the ground speed is computed from the formula

$$f_d = \frac{4V}{\lambda} \cos\theta \cos\phi \left(\text{or, } \frac{4Vf}{c} \cos\theta \cos\phi\right) \tag{5}$$

using only one pair of Janus aerials i.e. 1 and 3 (or 2 and 4).

Notice that in the above formula, *2V* of formula 4 has increased to 4*V*. This is because a four-beam system is being used and the responses from the front (F) and rear (R) beams of one of these pairs are compared directly and not with the transmitter frequency. Mathematically:

$$\text{front response: } f_{R,F} = f + \frac{2V\cos\theta}{\lambda}$$

$$\text{rear response: } f_{R,R} = f - \frac{2V\cos\theta}{\lambda}$$

$$\text{difference } f_{R,F} - f_{R,R} = \frac{4V\cos\theta}{\lambda}$$

Note: while a Janus four-beam system obtains *ground speed* by measuring the Doppler beat frequency from a single pair (front/rear) of Janus aerials, i.e. utilises only two aerials, *drift* is obtained by comparing the beat frequencies from both pairs of Janus aerials, i.e. utilises all four aerials.

Janus aerials

A name given to the aerials which transmit forward as well as rearward, and comes from the Roman god who could simultaneously look forward and behind. These aerials are used with advantage in various ways.

(1) To reduce errors due to the transmitter frequency wandering. Accurate measurement of frequency differences depends largely on the size of such differences. To make them large enough (about 10 kHz), Doppler equipment transmits at a very high frequency. An airborne transmitter working at these high frequencies is unstable and may wander.

With systems which require the reflected energy to be compared with the transmission frequency, the only transmission frequency that is available for the comparison is the frequency actually being transmitted at that instant. If this is not the same as the frequency to which the reflection relates, an error in the beat note will occur.

With the Janus system this coherence is not critical because signals are transmitted simultaneously forward and backward. The signals return to the receiver simultaneously. These two signals contain a component of the transmission frequency f in them and therefore, when mixed, a beat frequency is extracted correctly. With the pulse system the transmission errors may be completely ignored and with CW systems, they can be compensated.

(2) To reduce pitch errors. In a single-ended system, an error is introduced with the change in pitch angle. For example, when the aerial is pitched up $p°$, the Doppler frequency becomes

$$\frac{2V\cos(\theta - p)}{\lambda}, \text{ see Fig. 16.6.}$$

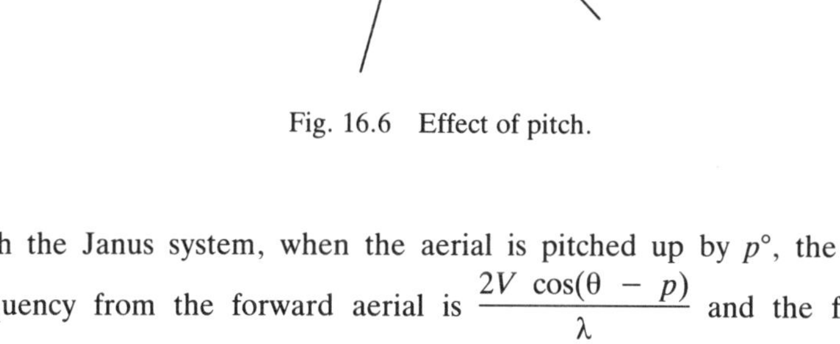

Fig. 16.6 Effect of pitch.

With the Janus system, when the aerial is pitched up by $p°$, the Doppler frequency from the forward aerial is $\frac{2V\cos(\theta - p)}{\lambda}$ and the frequency from the backward aerial is $\frac{2V\cos(\theta + p)}{\lambda}$. As these two frequencies are mixed to extract the beat note and around a depression angle θ of 67° changes of cos θ are nearly linear for small values of *p*, pitch errors are significantly reduced even if not completely eliminated.

(3) A Janus aerial, by producing a conical beam pattern ensures an overlap period between two signals by prolonging in time. Thus the equipment will not unlock when flying over uneven territory.
(4) With this type of beam, when the aircraft rolls, it does so on the edge of the beam and the returned frequencies are not affected.

Doppler spectrum

Although the Doppler equipment transmits signals at a given depression angle, because of the beamwidth, an area of the surface rather than a single point is illuminated (Fig. 16.7). The beamwidth varies between 1° to 5° depending on the type of the equipment and because of this, the reflected signals are not composed of a single frequency but a spectrum of frequencies. In Fig. 16.8, the angle of depression is assumed to be 60° and the beamwidth is 4°. The polar diagram of the spectrum, that is, the shape of the curve, is determined by the aerial design. In this particular illustration the energy

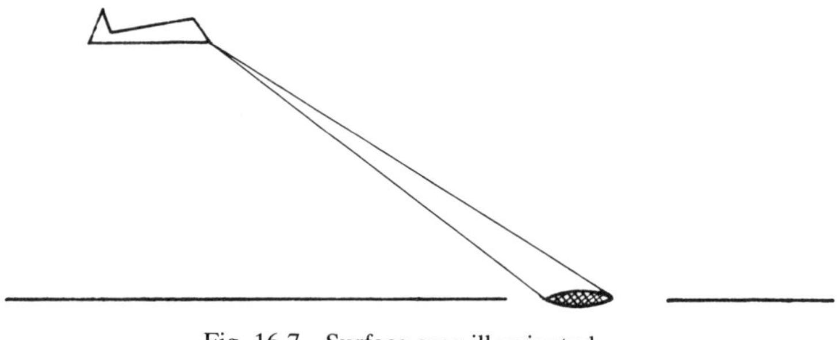

Fig. 16.7 Surface area illuminated.

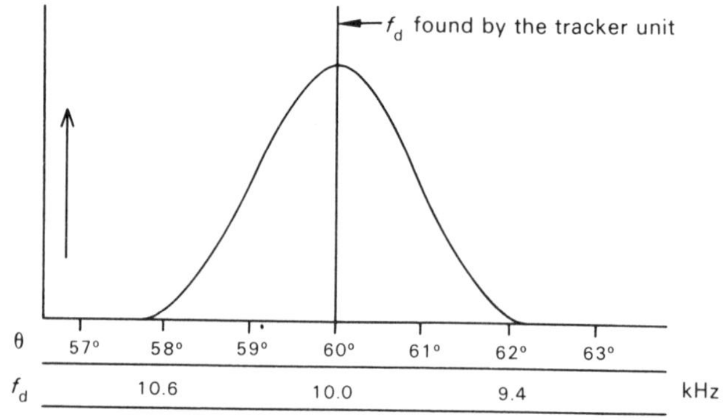

Fig. 16.8 Doppler spectrum.

envelope consists of frequencies varying from $\frac{2V \cos 62}{\lambda}$ to $\frac{2V \cos 58}{\lambda}$. A unit in the receiver called the frequency tracking unit finds the f_d by dividing the energy content envelope in two equal halves.

In practice, the shape of the curve is not symmetrical because more energy is reflected away from and less energy returned to the aircraft from those parts of the beam furthest away from the aircraft. The smoother the surface, the greater this effect will be, especially over the sea. The derived mean frequency and hence the measured ground speed will fluctuate slightly as the aircraft passes over surfaces having differing reflectivity.

Airborne Doppler

Transmitter/receiver

The Doppler transmission takes place on two frequency bands: 8800 MHz and 13 500 MHz. The use of such high frequencies is necessary for two reasons:

(1) The transmission beamwidth must be as narrow as possible in order to give a narrow Doppler spectrum. A narrow beam can only be radiated at micro wavelengths.
(2) The higher the frequency shift arriving at the receiver, the greater the accuracy of the measurements and the result. Higher transmission frequencies produce higher frequency shifts. The use of higher frequencies is, however, accompanied by the inherent disadvantage of being subject to increasing atmospheric attenuation. Therefore, the ultimate operating frequency is decided from the consideration of the power output available and the operational height required.

The receiver, having carried out a search for the returning signals and having accepted them, locks on to them. It delivers the signals to the tracking unit in the form of a Doppler beat frequency spectrum.

Tracking unit

The tracking unit accepts the frequency spectrum from the receiver and finds the mean frequency. It uses this to measure the aircraft's ground speed. It also controls the rotational movement of the aerial. The operation of the tracking unit is shown in block schematic form in Fig. 16.9.

Where the aircraft is fitted with a compact navigation computer unit, this will process the inputs from the four-beam or three-beam system within the single unit and output the data to the indicator unit.

Indicators

Drift and GS indicators. These indications may be either by pointer-type indicators, digital or CRT displays or a combination of them. Indicators used

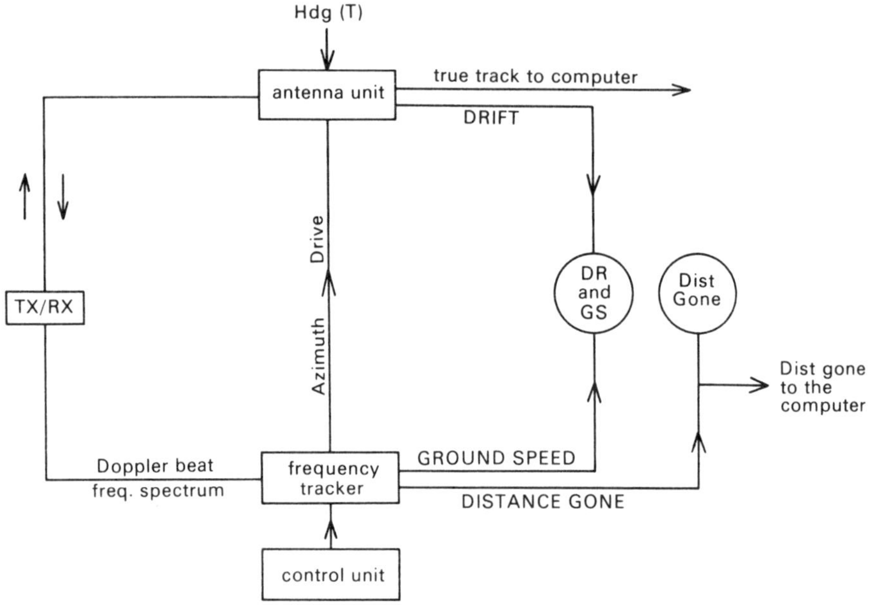

Fig. 16.9 Doppler tracking unit.

in early RACAL Doppler systems are shown in Fig. 16.10a but these have been superseded, first by the position bearing and distance indicators using LED displays (Fig. 16.10b), and subsequently by their CRT-type CDUs. (Fig. 16.11)

With some early systems, before the equipment would lock on to the correct values of drift and ground speed the indicator pointers had to be set manually, by means of 'inching' switches to the approximate values of expected drift and ground speed. Thereafter, the pointer would pick up the position to indicate correct drift and ground speed automatically, and any changes due to WV or alteration of headings would be followed.

Distance to go/distance gone indicators. The facility to obtain alternative presentation is sometimes incorporated in the equipment; otherwise you express your choice when buying the equipment. The counters are initially set to read either 000 or the leg distance. Alternatively in navigation management systems, waypoints are entered which can become the reference points from which distances can be expressed. In flight, the indicators are operated by electrical impulses originated in the tracking unit or airborne computer.

The ground speed indicated on the GS meter is a spot GS and therefore, where a track plot is being maintained, the ground speed should be calculated from the readings taken from the distance gone meter over a period of time.

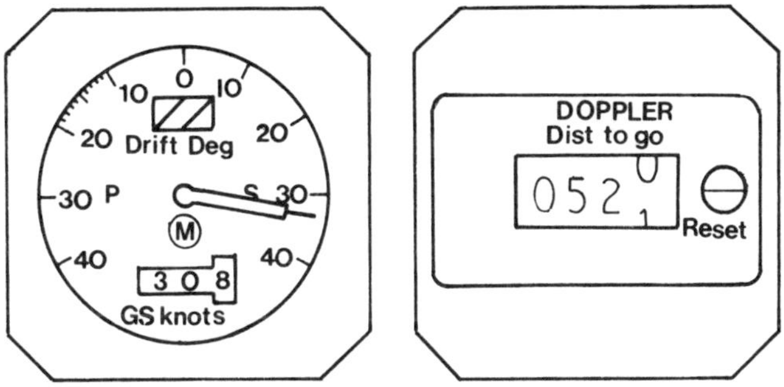

(a) early indicators

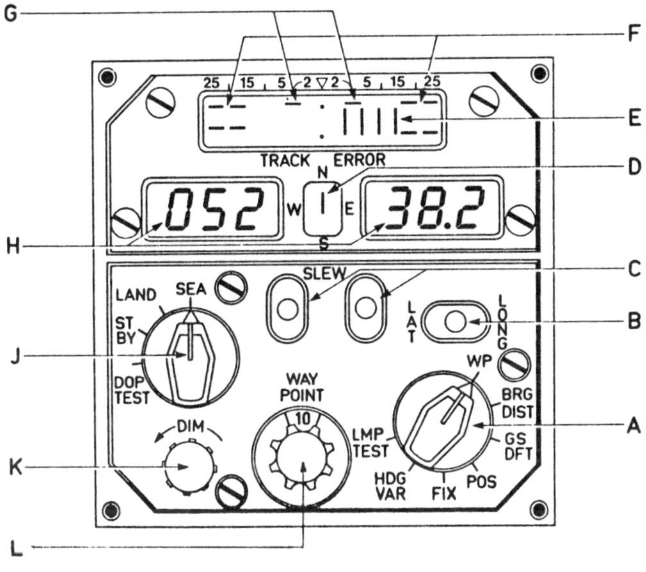

A: Display switch. B: Lat/long selection switch. C: Slew switches (centre-biased and used to update data). D: Light bars to indicate E/W variation, N/S lat, E/W long, etc. E: Track error indicator. The more bars on the right the more the heading must be altered to the right to regain the required track. F: Warning bars (flash to draw pilot's attention). G: Power supply test bars (used with J). H: Numeric displays. J: Function switch. K: Display illumination brilliance control. L: Waypoint selection switch.

(b) position bearing and distance indicator

Fig. 16.10 Doppler displays.

Memory facility

Most equipment incorporates a memory circuit which takes over when the received signals are weak, that is, below a predetermined value and the receiver unlocks. In these circumstances, the drift and ground speed pointers remain locked to the last strongly measured signals. When strong signals are received again the pointers are released. If drift and ground speed information is being used to give position indications, the resultant position

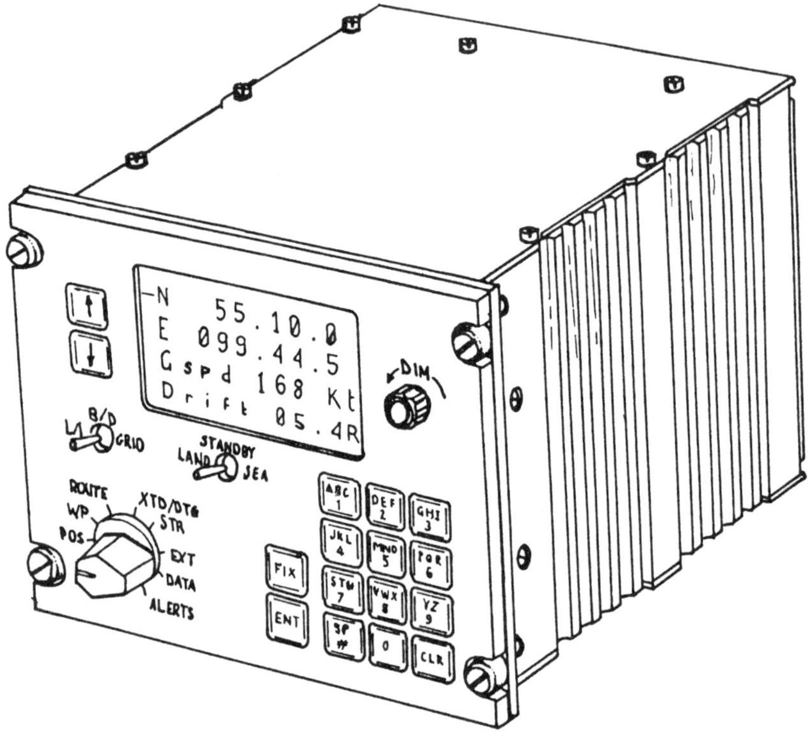

Fig. 16.11 Navigation computer presentation of Doppler position ground speed and drift.

information is only as accurate as the 'frozen' drift and GS shown on the meters. Under these circumstances the position information is only equivalent to the dead reckoning (DR) plot and when the strong signals return, it must be updated if you have been running it on memory for a long time or substantial changes in the WV are suspected.

Doppler limitations

As mentioned above, Doppler will unlock when the signals being received are weak. It will also unlock and go to memory mode in the following circumstances.

(1) *When flying over calm sea.* Calm sea reflections are not sufficiently scattered to give strong echoes back. It needs at least a 5 kt wind to produce sufficient sea surface irregularities for satisfactory operation.
(2) *Atmospheric conditions.* The receiver will unlock if severe thunderstorms are present around the aircraft. The outgoing signals will be reflected by the water drops and admit radio noise whereas the reflected signals will be weakened due to attenutaion.
(3) *When flying very low.* The Doppler receiver is short-circuited when

transmission is taking place. At very low levels, by the time the receiver circuit is in operation again the reflections will have passed the aircraft.

(4) *Limits in pitch and roll.* If the limits in pitch and roll as given by the equipment manufacturer are exceeded, Doppler will unlock and the 'memory' circuit is activated. Typical limits are ±20° in pitch and ±30° in roll.

(5) *Height hole effect.* For a pulse transmission system, this occurs when the aircraft is at such a height that the time taken for a signal to reach the ground and return is equal to the time interval between pulses, or to a multiple of that time. While a pulse is being transmitted, the receiver is switched off, and the reflections are not received.

This effect is not usually a prolonged one when flying over a landmass because of the surface irregularities. For pulse radar, the effect is avoided by gradually changing its frequency; for CW radar, by further frequency modulating the signal.

Indicators and associated equipment

Apart from displaying the Doppler's basic information, ground speed and drift, it can be used in conjunction with the true heading reference from the aircraft's remote indicating compass to provide position information in a variety of ways:

- latitude and longitude
- bearing and distance from a selected waypoint
- distance flown along track and across track
- position in terms of a square grid.

With the more sophisticated systems, WV, required track and GS, ETA, steering information, etc., may be provided and the Doppler data may be used to run a moving map display. Typical indicators have evolved from the early displays of Fig. 16.10(a) to the later Figs 16.10(b) and 16.11 and the navigation display on the EFIS.

Latitude and longitude. Using the ground nautical miles flown and the TMG information, the airborne computer solves the triangle shown in Fig. 16.12 to display latitude and longitude of the aircraft.

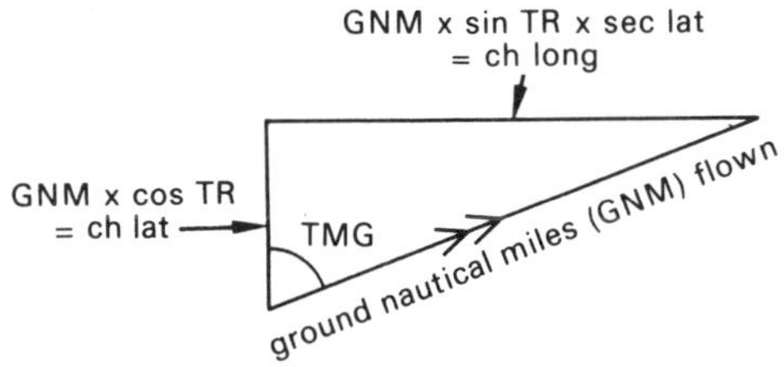

Fig. 16.12 Determining position.

Errors of Doppler equipment

The accuracy of the basic information is very high: generally, a GS accuracy of 0.1% and a drift angle accuracy of 0.15% may be expected, while maximum error (on 95% of occasions) is not likely to exceed 0.5% in GS and drift measurement. Errors arise from:

- aerial misalignment
- pitch or vertical speed error
- sea movement error
- sea bias error – Doppler return is distorted
- errors input from associated equipments (heading, latitude, altitude).

The greatest error of a Doppler system is due to its reliance on the accuracy of the heading input information, which when the Doppler drift is applied to it can give a considerable accumulative error.

Error effects may be summarised as shown in Table 16.1:

Table 16.1

Factor	Error in the measurement of distance	track
Aerial misalignment	No	Yes
Pitch error	Yes	No
Sea movement	Vector error dependent upon time spent over tidal water, direction of movement of water and the wind velocity	
Sea bias	Yes	No
Heading error	No	Yes
Altitude and latitude	Yes (affects latitude and longitude information)	No

Updating Doppler

The fact that Doppler may unlock on occasions and go into memory has already been mentioned. Resultant errors together with the system errors and errors in climb/descent accumulate with time. If during flight a better fix is available at any time, it is compared with the information being indicated from the Doppler and if there is any discrepancy, the Doppler co-ordinates can be brought up to date by the computer. This process of adjusting the co-ordinates to the latest and accurate fix is called updating Doppler. The fix being used for updating must be of a high accuracy, e.g. Decca, global positioning system (GPS) or derived from two simultaneous DME ranges.

In spite of all these errors of Doppler, when flying at a high level a pinpoint beneath the aircraft is not considered sufficiently accurate to be used for updating Doppler.

Advantages of Doppler

- It is an independent aid
- It has no range limitations
- It can be used world-wide
- It is of high accuracy
- It has area navigation capability.

Other aviation uses of the Doppler principle

(1) ***Doppler DF.*** If a ground station's vertical aerial receiving an aircraft's transmission is moved rapidly, this will also produce a Doppler effect. If the aerial is moved towards the aircraft there is a rise in frequency and if it is moved away from the aircraft, there is a received frequency fall.

If the DF station aerial is mounted on a long radial arm and the aerial is moved around a circular path, the frequency of the incoming signal at an aircraft will vary sinusoidally. For a constant rate of aerial travel, the incoming signal will be frequency modulated (at the frequency of the aerial rotation) and the phase dependent on the aircraft's bearing from the ground DF station. The phase can be measured against a fixed reference signal. Unfortunately however, the radial arm on which the DF station aerial would have to be mounted would in practice have to be some 5 m in length and rotating around its hub at up to 20 times per second, which is not particularly practical! Nevertheless, the same rotating aerial effect can be produced by an alternative means. This is by using a ring of vertical static aerials from which the received frequency is sampled in turn, getting the same result as would be obtained by the moving aerial. The phase comparison is then made and the direction of the incoming Doppler-affected signal displayed (to an accuracy of ±1°) on a CRT, LCD or mechanical meter.

(2) ***Doppler VOR***. Although already described in general terms in chapter 6, it is appropriate to refer here to the Doppler principle involved. In fact the procedure employed is very similar to that just described under Doppler DF except that in this case it is the ground transmitter aerial which is moving instead of the ground receiver aerial.

The central DVOR aerial generates an omnidirectional signal amplitude-modulated at 30 Hz. The circular pattern of DVOR aerials is energised in turn 30 times a second, with a separate continuous wave transmission 9960 Hz displaced from that energising the central aerial, by means of either a solid state switching device or by a commutator, to simulate the rotation of a single aerial. When the antenna appears to move towards the aircraft the frequency increases and as it appears to move away the frequency decreases. The combination of a circle of a diameter of 44 ft and a rotational speed of 30 revolutions per second gives a radial velocity of about 4150 ft/sec. This in turn will cause a maximum Doppler shift of 480 Hz which matches the

VOR requirements. The 9960 Hz frequency difference is therefore varied by ±480 Hz at 30 Hz rate with the phase dependent upon the bearing of the aircraft. The aircraft receiver is therefore receiving from the DVOR the same information as from a conventional VOR.

Chapter 17
HYPERBOLIC NAVIGATION PRINCIPLES: LORAN-C

Principle of hyperbolic aids

Hyperbolic navigation aids have been available to aviators for over 50 years, from the days of the Gee radar chains used by the RAF in the Second World War. A hyperbola is defined as the locus of a point having a fixed difference in range from two other fixed points. In the earlier types of navigation aids, hyperbolae were pre-printed on special charts for that aid and position lines obtained from the aid were plotted by interpolation between the hyperbolae. Nowadays our flight-deck equipment enables simultaneous position lines to be obtained, the fix determined and the position displayed on the pilot's instrument panel without recourse to the hyperbolic chart at all. Irrespective of the type of radio navigation aid, consider the principle illustrated in Fig. 17.1. Here M and S are the two fixed points, representing in this case a master station and slave station.

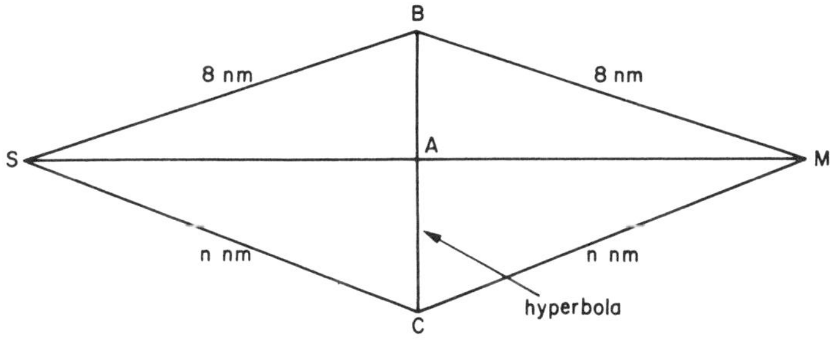

Fig. 17.1 Perpendicular bisector/hyperbola.

Join M–S by a straight line. The straight line thus drawn is called the base line between the two fixed points. BAC is the perpendicular bisector of MS.

Point A is halfway between M and S and, say, the distances A–M and A–S are 5 nm respectively. The difference between the two ranges = 5 − 5 = 0.

Point B is 8 nm from M. Therefore, it must be 8 nm from S, BAC being the perpendicular bisector of MS. Differential range equal to 8 − 8 or 0.

If point C is 'n' nm from both M and S, its differential range is also 0.

Draw up a smooth curve (it will be a straight line in this case) BAC which is a hyperbola. The curve BAC may be extended in both directions and any point on that curve will yield zero differential.

Curve BAC would not be the only hyperbola between M and S. Any number of hyperbolae may be constructed to cover the area. Take point P, a distance of 7 nm from M and 3 nm from S, along the base line (Fig. 17.2).

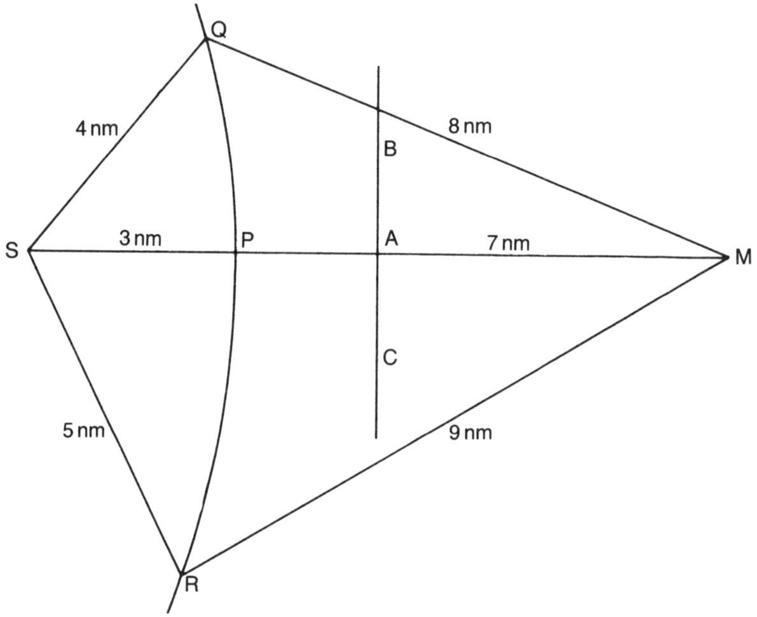

Fig. 17.2 Hyperbola QPR of d = 4.

Point Q is 8 nm from M and 4 from S, giving a differential range of 4.
Point P is 7 nm from M and 3 from S, differential range 4.
Point R is 9 nm from M and 5 from S, differential range 4.

Curve QPR is a hyperbola of differential range of 4 nm.

To utilise the system we want equipment in the aircraft which measures the differential ranges. Then, once we know that the differential is 4, for example, we know that we are on curve QPR, a position line. But to plot ourselves on that curve we also need printed hyperbolic charts. Such charts are available for use in conjunction with Loran or any other existing hyperbolic system.

As for the information, it is not necessarily needed in terms of nautical miles as in the above illustrations. It may well be in terms of phases of the signal received in the aerial (Omega and Decca) or in the time in microseconds (μs) that the radio wave took to travel to the aircraft (Loran). In either case, the principle still remains the same, that of differential range. For example, if the information is in terms of time, at point R, signal from M will take 55.62 μs (6.18 μs = 1 nm) and the signal from S will take 30.9 μs to arrive at the aircraft. The differential of 24.72 μs will be labelled against the hyperbola QPR instead of 4 nm as in Fig. 17.2.

However, as has already been mentioned, we have moved on from the days when position lines were first plotted on hyperbolic lattice charts on the flight-deck then to be plotted as a fix on a Mercator Chart and compared to the airplot to navigate our aeroplanes. Modern technology has relieved us of this chore and pilots now enjoy a continuous display of position derived from the hyperbolic aid, directly on the flight-deck panel.

Loran-C

Loran is a hyperbolic navigation system, deriving its name from 'Long Range Aid to Navigation' and was originally designed to cover the open oceans, having a range of up to 2000 nm.

Loran-C works on LF (100 kHz) on the principle of differential range by pulse technique. Transmitting stations mainly operated by the US coastguard are arranged in 'chains', each chain comprising a master station (M) with two or more secondary stations or slaves (S) arranged around the master. There are wide variations in the length of the baselines between the master and slaves of a chain depending upon requirements and suitable locations, but the maximum baseline length is about 1000 nm. The slaves are designated Z, Y, X and, where there are four slaves, W.

If both the master and slave in a chain transmitted synchronously it would be impossible to determine which was which. Also, it is necessary to be able to identify stations in different chains when within the coverage of more than one chain. In the original Loran of the 1940s, single pulses were transmitted and slaves operated on different frequencies with particular allocated pulse recurrence frequencies (PRF). Now Loran-C signals are transmitted in groups of pulses. The pulse groups recur at a fixed group repetition interval (GRI), the value of which is different for each chain (and designated by a code). In order to achieve the greatest efficiency, the time difference is read from the third cycle within a group of pulses from the master and from the chosen slave (Fig. 17.3). (This process is also known as 'indexing'.)

The master station in a chain transmits a group of pulses in all directions. Its slave station receives this signal and the slave's transmitter is triggered but delays transmission for a specific time interval called the coding delay. It then transmits in all directions, the coding delay having been long enough to ensure that the slave's signal could not arrive at the aircraft before the master signal, whatever the aircraft's position in the chain's coverage area.

With the first Loran sets the operator, having tuned in a chain, measured the time difference between and M and S signals on a CRT screen. This was a complicated process requiring experience and dexterity in identifying the master and slave signals, switching the signals to a double timebase where they had to be balanced, reading their time difference from a superimposed calibration scale and, if necessary, making ground wave/sky wave correction

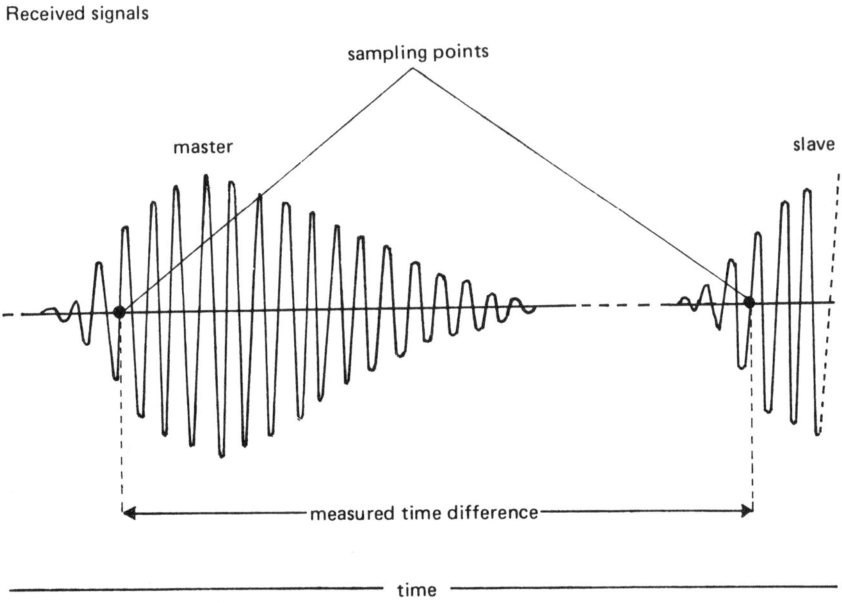

Fig. 17.3 Indexing.

before the position line could be plotted on a Loran chart. Such a Loran chart for a particular area usually bore the hyperbolic lattice for two or three chains, overprinted in a different colour to avoid confusion. Often the hyperbolae were shown at 100 μs intervals and interpolation was necessary for intermediate values.

The advent of the airborne computer and multi-chain receivers has enabled the modern Loran-C system to have a direct and continuous display on the flight-deck of the aircraft's position in latitude and longitude, derived from the Loran signals it is receiving. It can be coupled with a navigation system containing predetermined waypoints of a stored flight plan. On departure the pilot turns on the equipment, selects a pre-stored route, and the system then operates automatically without any further pilot intervention. The chains and stations are acquired and deleted as necessary between the successive waypoints. Some equipments are capable of tracking up to eight stations in four different chains simultaneously and using all of these stations in the navigation solution instead of being restricted to three stations in a single chain.

On the modern flight-deck, Loran-C also usually offers alternative navigational data to be presented against the waypoints which have been either stored pre-flight or set up in-flight in the navigation system, e.g. as bearing and distance, track and track error, etc.

Depending upon the CDU on the flight-deck, the pilot may be presented with either the time difference between specific stations to a resolution of 0.1 μs or, more usually nowadays when used in conjunction with an airborne computer and selected on the CDU, the aircraft's position in latitude and longitude or relative to chosen waypoints.

Normally the accuracy of Loran-C is in the order of 0.1 to 0.2 nm in areas of good cover, decreasing to 0.5 to 1 nm at 1000 nm. When the CDU is displaying time-difference for the master and two slaves, it is desirable to get the best angle of cut between the position lines. In these circumstances it may be necessary to discard one of the slave signals being tracked and acquire another slave which will give a bigger angle of cut.

Most modern Loran CDUs have 'status' indicators which enable pilots to have a visual indication of the quality of signal reception. In the event of a malfunction or irregularity originating at a Loran-C chain, the stations of that chain transmit warning signals. These actuate a visual indication on the CDU, e.g. 'blinking' signals. Similarly, the CDU is usually provided with BITE which enables overall system performance to be checked without external equipment, on the ground or in-flight, whether or not Loran signals are available.

Test questions

(1) Loran is:
- (a) the name of the inventor of the system
- (b) an acronym of Longwave Oceanic Radio Aid to Navigation
- (c) derived from Long Range Aid to Navigation.

(2) Loran is available for use:
- (a) courtesy of the US Department of Defense
- (b) in N America, N Atlantic, Europe and Mediterranean where there are chains
- (c) world-wide pole to pole.

(3) Loran operates on:
(a) 100 kHz (b) 1000 kHz (c) 106 kHz.

(4) Loran position lines/fixes in the coverage area are:
- (a) available both day and night
- (b) are unreliable at morning and evening twilight
- (c) are unreliable along the chain baseline.

(5) The time difference is measured in a Loran receiver by a process known as:
(a) scalloping (b) indexing (c) hyperbolicking.

Chapter 18
DECCA

Decca is another hyperbolic navigation system, operating on the master–slave basis. It differs from Loran-C in many ways. First of all, the basic principles employed by the two systems to produce differential ranges are different. Decca measures differential ranges by comparison of phases of the master and slave signals arriving in the aircraft. The presentation of information may be different – Decca originally displayed information by means of three decometers, on the dial and pointer system, although it is now usually presented as navigational information on a CDU. Decca operates on yet lower frequencies, the LF band in fact, and lastly, Decca is a short range navigation aid.

A standard Decca chain consists of a master transmitter and three slave transmitters, the slaves being known as red, green and purple. The slaves are placed around the master, approximately 120° apart from each other and at a distance of between 40 and 110 nm from the master. One master–slave pair gives one position line and information from all three pairs is continuously acquired to give the aircraft's position.

Principle

The principle of Decca is differential range by phase comparison. Let us see first of all, how it is possible to have a knowledge of range simply by measuring the phases of two signals and comparing them. See Fig. 18.1.

It is an M–S combination, each transmitting at two cycles per second. At any given instant, the signals relationship is as shown in the figure. The master's signal is shown as a continuous line from M as wave A, the slave's as the pecked line from S as wave B. Both waves are phased locked, that is, their crests and zero values occur at the same instant.

Let us now consider the phases produced by these two waves at different positions in the area MW. Signal A (master) will produce phases of 045, 090 and 135 respectively at positions a, b and c. At these same positions signal B (slave) will produce phases, 315, 270 and 225. Both signals will produce 0 phases at M and 180 at W.

Now, if we had a meter sensitive to phases, at position M it would read 0 phase from both the signals. Further, if this meter was capable of displaying the result as a difference of the two phases, it would still read 0, as shown by a meter in the figure. If we now move to position a, the phase meter will read

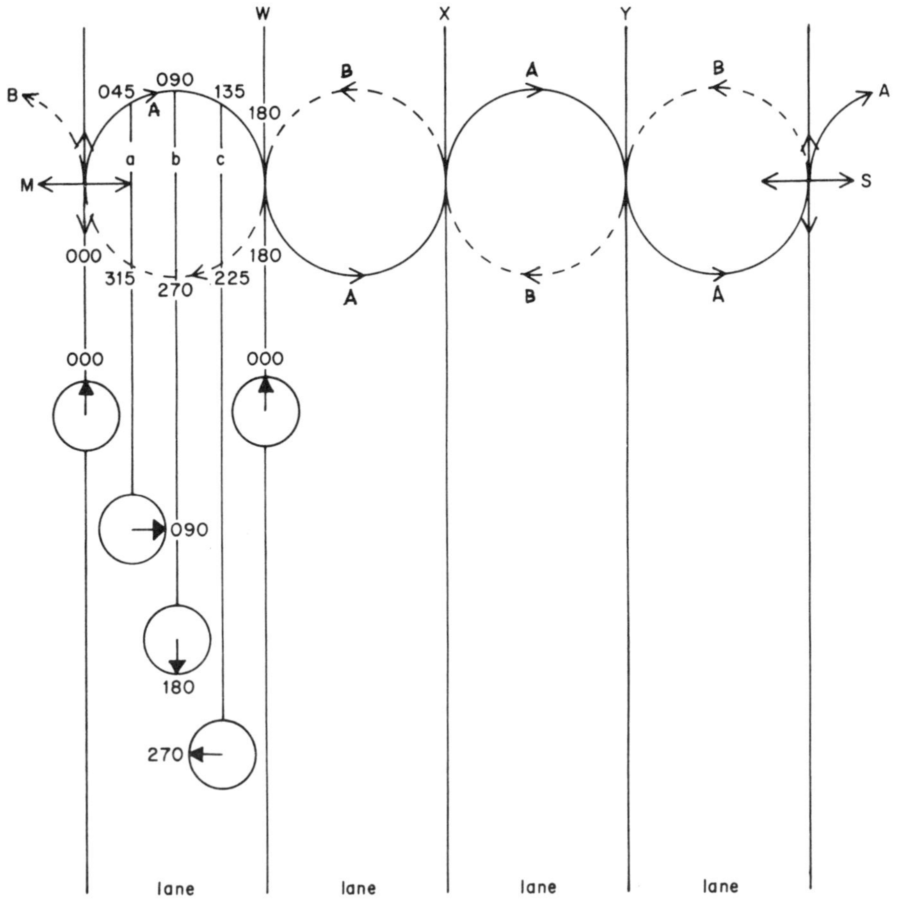

Fig. 18.1 Master and slave phases.

a phase of 045 from A signal, 315 from B signal, and the result displayed will be 045 (+360) − 315 = 090.

At b, the phase difference will be 090 (+360) − 270 = 180
At c, the phase difference will be 135 (+360) − 225 = 270

Lastly, at W, the phase difference will be 180 − 180 = 0.

Thus, as we moved from M to W, we went through an area of complete 360° phase difference, or the needle in Fig. 18.1 completed one full revolution. This is the basis of the principle. We can calculate the distance from M to a point where, for example, a phase difference of 090 or any other given value will occur. We can do this because the distance M–W is the distance of half the wavelength. We know the frequency, therefore, we know the wavelength.

In Fig. 18.2 a hyperbolic lattice is drawn up for a two-wave transmission as illustrated in Fig. 18.1.

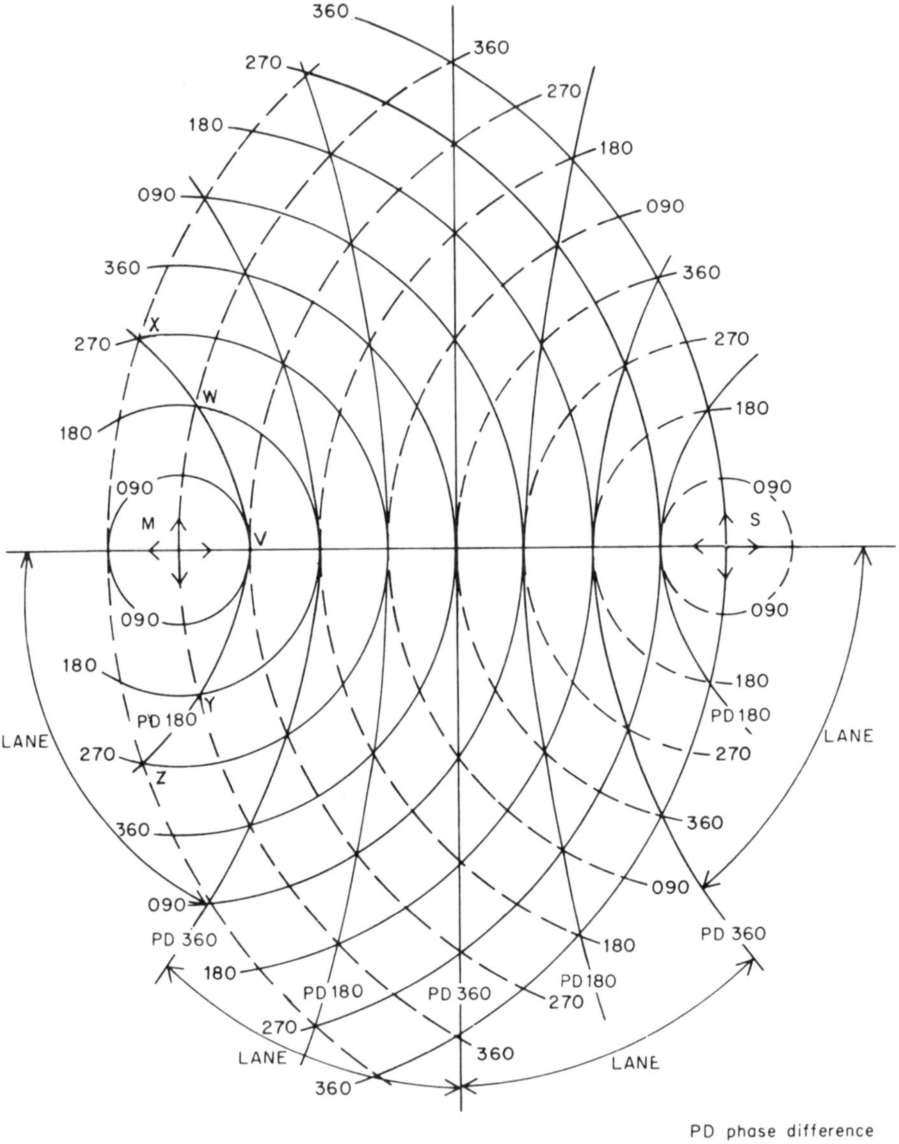

Fig. 18.2 Hyperbolic pattern.

In this figure, master and slave waves are shown every 90°, master's transmission being continuous curves, slave's pecked curves. Hyperbolae are determined as follows:

Starting at M, the phase due to master signal at point V is 90, the slave phase is 270. Difference is 090 + 360 − 270 = 180.

Now we want to find all other points in the vicinity of M which will give the same phase difference, i.e. 180. Take point W. Master's phase here is 180, slave's phase 360, difference 180. Points X, Y and Z are similarly found. A smooth curve joining these points gives a hyperbola of 180 phase difference. This means that an aircraft anywhere along that curve will read a phase difference of 180, and a small portion of that hyperbola in the vicinity of the aircraft's position is its position line.

By a similar process hyperbolae at any convenient interval may be drawn up between master and slave and then the original wave pattern erased. Note that the perpendicular bisector is a straight line curve. Note also that ambiguity exists behind master and slave.

We are now ready to see how this principle is implemented in practice. In our illustration in Fig. 18.1, we have both master and slave transmitting on the same frequency. In practice, this is not possible since two signals on the same frequency arriving at the aerial will merge and appear as a single voltage. Unless separate identity is maintained the phase comparison cannot take place. Therefore, a Decca master and the three slaves transmit on different frequencies. Each station (i.e. master and three slaves) has a basic or fundamental frequency called f. The value of f is always in the region of 14 kHz. Master and its slaves then transmit on fixed harmonics of that fundamental frequency. Harmonics are the multiples or divisions of a given frequency and they are easier to produce. These fixed harmonics in respect of the four transmitters are as follows, and they are valid for a Decca chain anywhere in the world:

Master – $6f$
Red – $8f$
Green – $9f$
Purple – $5f$

Just to repeat for the sake of emphasis, the multiples remain constant throughout, the value of f varies from chain to chain. A typical chain operating on a basic frequency of 14.2 kHz is shown in Fig. 18.3.

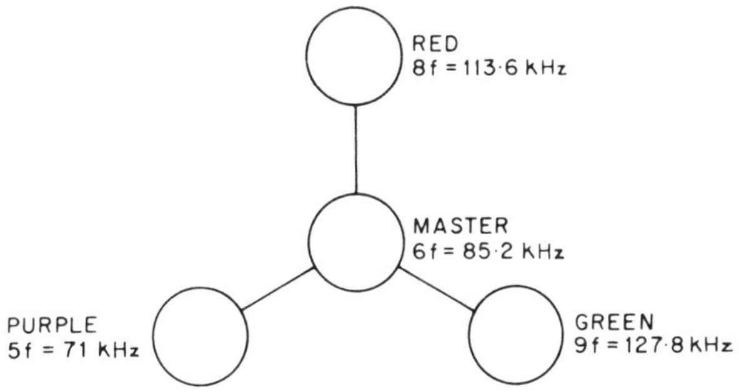

Fig. 18.3 Transmission frequencies.

Comparison frequency

By staggering the four transmission frequencies we solved the problem of keeping the signals separate. But in doing so, we created a new problem.

Phase comparison cannot take place between any two signals which are not on the same frequency for the simple reason that they do not bear phase relationship. Now we have two signals in the receiver but not on the same frequency and we cannot compare their phases. So where do we go from here? The obvious solution is to step them up now to a common frequency and then take the phase difference; Decca has done just that. Each master–slave combination is stepped up in the receiver to the value of their lowest common multiple (LCM) and phase comparison is then taken. These stepped up values are called comparison frequencies, and they are as follows:

Master and red

M transmits at $6f$
R transmits at $8f$ LCM = $24f$ which is the comparison frequency

Similarly, the comparison frequency for master and green is $18f$ and master and purple, $30f$.

Thus, the principle is finally implemented and the result of phase comparison could be displayed by a pointer on individual Decca meters called Decometers. The only observation that remains to be made is how the receiver converts the received frequencies into comparison frequencies. This is shown in Fig. 18.4 which is self explanatory.

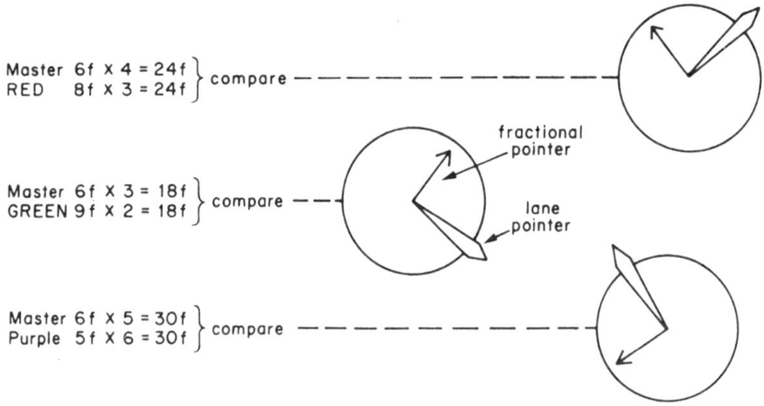

Fig. 18.4 Basic Decca information presented on decometers.

Decometers

With the information coming to the Decca receiver in the form of phase differences it is now necessary to create a system of presenting this information in a convenient form so that the operator can rapidly utilise it.

For a start, the very fact that a 360° phase change occurs over a known distance can be adopted to give us a basic unit of measurement. On a decometer (which is a Decca indicator) these 360° phase changes are indicated on the fractional pointer – Fig. 18.4. The pointer indicates the position inside any one of the numerous 360° phase change areas that occur between a master and a slave by taking the phase difference between the two signals.

The next step is to be able to locate the exact area we are in. To enable us to do this, these 360° phase change areas, called Decca lanes are numbered, the numbering system being explained below. A lane is defined as an area enclosed by two hyperbolae of 0° phase difference. On a decometer, the lane number is indicated by the lane pointer geared to the fractional pointer.

The physical distance that any one lane would occupy, that is its lane width, depends on the slave's comparison frequency, because it will be remembered that the fractional pointer is placed by the phase difference taking place at this frequency. Along the base line, the lane width is the distance occupied by half the wave at comparison frequency – see Fig. 18.1. Taking a typical Decca fundamental frequency, f, of 14.2 kHz, the lane width for the three slaves is calculated as follows:

Red slave: compares at $24f$, comparison frequency = 24 × 14.2 = 340.8 kHz.

$$\text{Lane width} = \frac{300\,000}{340.8 \times 2} = 440\,\text{m}$$

Green slave: compares at $18f$, comparison frequency = 18 × 14.2 = 255.6 kHz

$$\text{Lane width} = \frac{300\,000}{255.6 \times 2} = 587\,\text{m}$$

Purple slave: compares at $30f$, comparison frequency = 30 × 14.2 = 426 kHz

$$\text{Lane width} = \frac{300\,000}{426 \times 2} = 352\,\text{m}$$

From the above it will be noticed that the different slaves produce lanes of different widths, green for example is 587 m whereas the purple is merely 352 m. It is further pointed out that no two slave transmitters can be expected to be located at exactly the same distance from the master. In a Decca chain, the green slave may be, for example, 160 km from the master and the red may be 180 km. These variables call for some kind of grouping of the lanes for the sake of standardisation of the calibration of the decometers.

This standardisation is achieved by the establishment of Decca zones, so that a zone provides a fixed distance measurement for all three slaves. A zone is defined as an area enclosed by two 0 phase hyperbolae formed by comparison at the basic frequency, f. The basic frequency, f, being the same for all three slaves in a chain, the zone width for all three slaves will also be the same. Along the base line, the width of a zone is the distance occupied by half the wave at the fundamental frequency, f. Zone width for any slave may be calculated as follows:

$f = 14.2$ kHz

$$\text{zone width} = \frac{300\,000}{14.2 \times 2}\,\text{m}$$
$$= 10\,563\,\text{m}$$

Thus, the zone width is independent of the slave colour, but within these zones, the number of lanes enclosed for each different slave will be different, as shown below:

Red: lane width = 440 m and the number of red lanes in a red zone

$$= \frac{10\,563}{440} = 24$$

Green: lane width = 587 m and the number of green lanes in a green zone

$$= \frac{10\,563}{587} = 18, \text{ and lastly}$$

Purple: lane width = 352 m and the number of purple lanes in a purple zone

$$= \frac{10\,563}{352} = 30$$

The original Decca equipment employed three decometers, one for each slave and the information provided was in the form of a zone, lane and the fraction of the lane, as below.

Zones. All the decometers read up to ten zones. These were lettered from A to J (including I) and the indication was on moving counters. If in a particular chain the master–slave distance exceeded ten zones, the letters then repeated. It will be appreciated that this repetition was not likely to cause an ambiguity because, on average the first ten zones would occupy a distance of around 100 km along the base line (and more elsewhere) before the repetition would occur.

Lanes. As mentioned earlier, the number of lanes in a zone differs according to the slave, and different calibrations for different slaves may be expected. The lane pointer moves round the dial calibrated as follows:

Red decometer – 24 lanes to a zone, numbered 0 to 23 (incl)
Green decometer – 18 lanes to a zone, numbered 30 to 47 (incl)
Purple decometer – 30 lanes to a zone, numbered 50 to 79 (incl)

Fraction of lane. 360° travel of the fraction pointer round the dials was shown in decimals, that is, from 0 to 0.99 rather than as 360 phases. This made it easier to read and the dial required fewer calibration marks which enhanced the accuracy of the reading.

To recap, when the fractional pointer completed one full revolution, a flight through one lane was complete. The lane pointer which was geared to the fractional pointer would then indicate the next lane. When the lane pointer completed one full revolution, the aircraft would have flown through 24 red lanes or 18 green lanes or 30 purple lanes. The zone indicator which was geared to the lane pointer would now have moved to indicate the next zone. An operator took the zone reading first, this being the slowest moving counter. The number of the lane was then read, followed by the fraction inside the lane. A typical reading D 41.75 is illustrated in Fig. 18.5. D is the zone letter, 41 is the lane number and .75 is the position inside that lane. Or, the aircraft has progressed three-quarters of the way inside lane 41 of zone D.

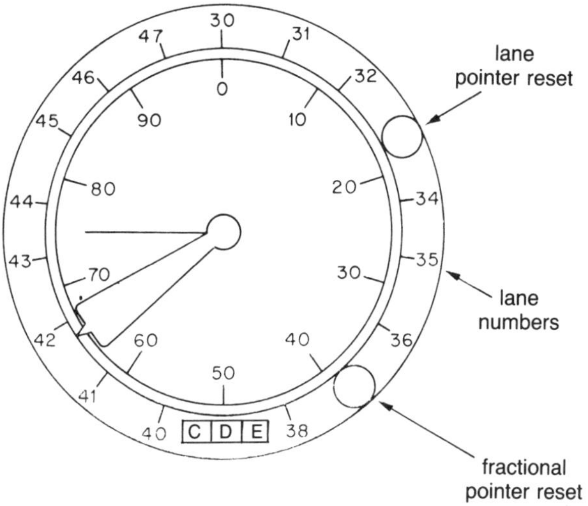

Fig. 18.5 Decometer reading D 41.75.

Lane identification

It will be appreciated that the length of a Decca lane is a very small distance indeed. For example, for a purple slave, if the master–slave distance is 100 nm, (and we know that there are 300 lanes in that distance) the length of a lane is a mere 1/3rd of a nm. If you are taking off from an aerodrome and planning to use Decca straightaway, this causes no problem, since you will have set the base co-ordinates in terms of zone and lane (fractional pointer will pick up correct position automatically) before the start. In that case, Decca will continue to indicate the correct reading throughout the flight. The problem arises in cases where temporary failure of the equipment occurs or when entering a Decca chain from outward or when changing over from one chain to another. In these instances accurate information on lane number is required. On the basic Decca, this information was provided on a lane identification (LI) meter, a single indicator calibrated in decimals just like the fractional pointers of the three decometers and which catered for all three slaves. The identification meter consisted of two pointers, one was wide and was called the sector pointer. The other was a six legged one and was called the vernier pointer (Fig. 18.6).

In the above illustration the reading is .72. If this was a green reading (that is, a reading taken when green light was on) and we wished to set the lane on

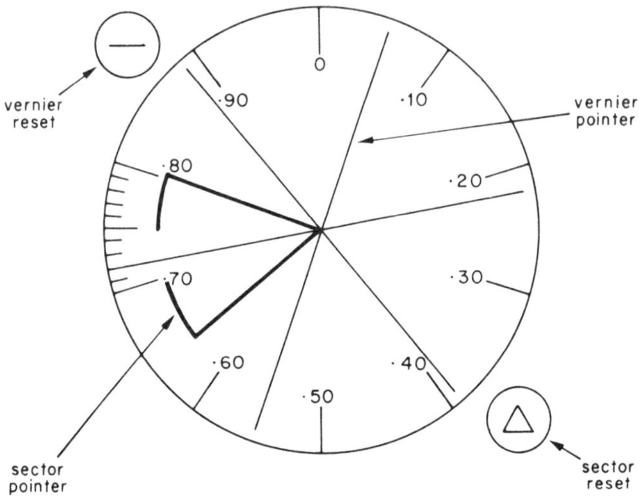

Fig. 18.6 LI meter.

the decometer in Fig. 18.5, we would move the lane pointer to cover .72 on fractional scale. This would put us in lane 43.

As for zone setting, it is wide enough to be ascertained by DR navigation. On later Decca models zone identification also takes place automatically.

We shall now look at the theory and see how LI is achieved, and particularly, how the lane number ties up with decimals on the fractional scale. Let us forget for a minute the LI meter and concentrate on the decometer in Fig. 18.5. As pointed out, the need for LI arises because of the smallness of the lane distances. The obvious answer to this would be to momentarily widen the lanes. In Decca the lanes are widened momentarily for LI purposes by providing suitable ground transmission to phase-compare the signals at $1f$ and $6f$. Fig. 18.7 represents a master and green comparison pattern. 180

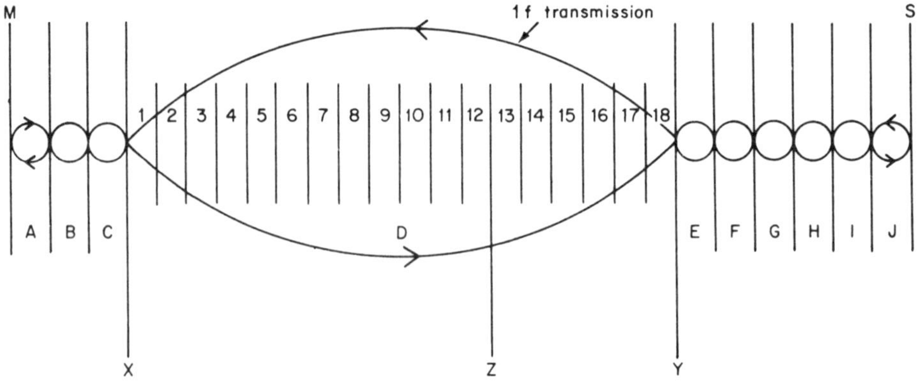

Fig. 18.7 Exploded view.

lanes are produced which are contained in 10 zones, each zone having 18 lanes in it. This is so during normal transmission. At $1f$, we know from earlier calculations, that there will be 5 waves between master and green (distance 60 nm) or 10 lanes. This $1f$ transmission forming 10 lanes between master and green is also shown in Fig. 18.7. Distance X–Y which marks the boundary of D zone (shown exploded for clarity) is now covered by half a wave from master and half from slave. This resembles the wave pattern between M and W in Fig. 18.1. Thus, as the aircraft travels from X to Y, 360° phase change will occur and the fractional pointer will complete one revolution.

Under this arrangement the distance X–Y (or original zone D) is a lane, and the fractional pointer at any time indicates position inside this lane in terms of decimals.

For example, when the aircraft is at Z, the fractional pointer of the decometer will have gone round 2/3rds of the dial, say, it indicates .72. Now, if against position .72 on the fractional scale, the decometer was calibrated to give lane numbers under normal comparison frequency, the reading would be 43 (30 + 13) from Fig. 18.7. If you examine Fig. 18.5, it is calibrated just that way. Fractional value at $1f$ indicates lane number at normal comparison frequency. As for the fractional pointer on the decometer, it would be most inconvenient if LI readings were fed to it. The needle would jump periodically to indicate a lane. Hence the use of a separate LI meter. The result of $1f$ is displayed on the sector pointer.

Comparison for lane identification also takes place at $6f$ simultaneously with $1f$. The reason for $6f$ comparison is to improve the accuracy of the LI. If you examine the $1f$ curve above you will notice that the curve is very gentle and the phases change very slowly. A comparison on its own would only yield an approximate result, and this accounts for the shape of the sector pointer. $6f$ will produce six lanes in the original D zone ($6f$ = 60 lanes in place of original 180. Therefore, the original three lanes will equal one new lane or, there will be six lanes in one original zone). So there will be six positions in distance X–Y where the same phase difference will be measured – hence the shape of vernier pointer. The leg that falls in the area covered by the sector pointer indicates the lane in decimals.

To enable the receiver to carry out phase comparisons at $1f$ and $6f$ the transmission pattern has to be modified.

Up to the late 1950s, with the early Decca system, for the purpose of lane identification the normal transmission was interrupted every minute. Since then, a multipulse (MP) type of LI has been in use and derives the required $1f$ signal from each station by a method in which twice as much information is transmitted as in the previous mode.

The complete transmission sequence of an MP chain is shown in Fig. 18.8.

Every 20 seconds the stations transmit the MP signals in the order MRGP together with an $8.2f$ component (for chain control and surveillance). The MP signals last 0.45 seconds and are spaced at 2.5-second intervals.

The interrupted and re-grouped chain transmissions enable the receiver to

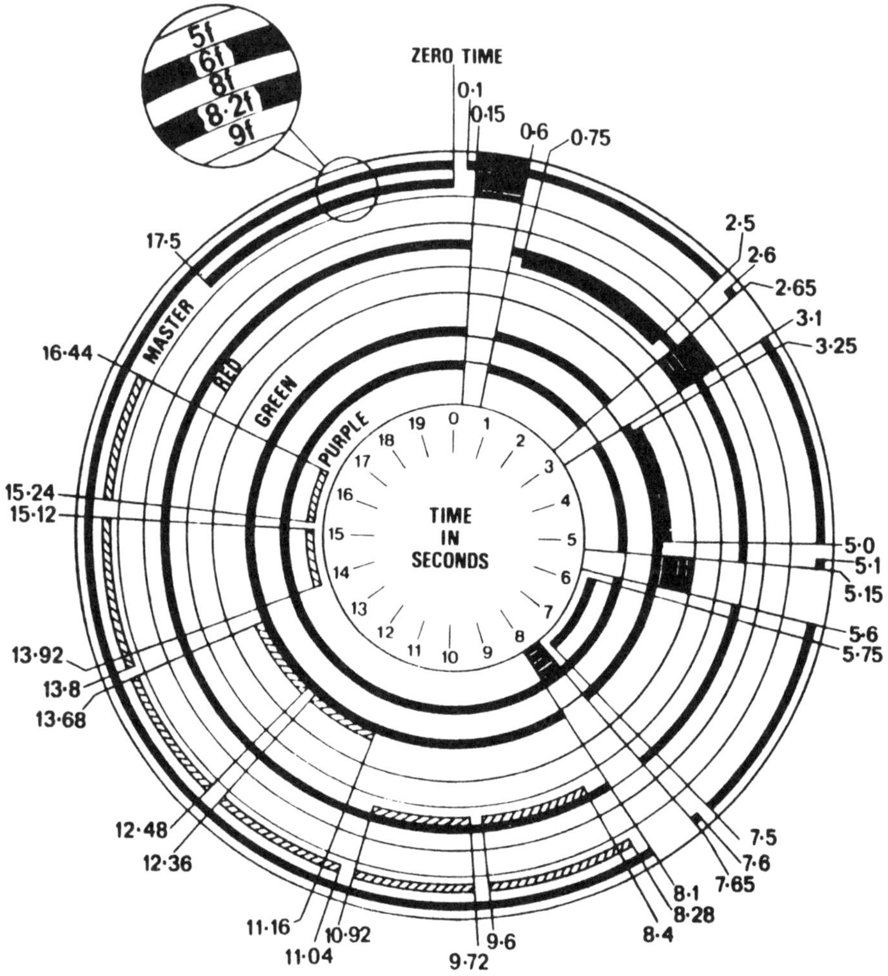

Fig. 18.8 Transmission pattern.

extract a signal of frequency f from the master and from each slave. Comparing the phase of these signals generates a coarse hyperbolic pattern cofocal with the fine pattern such that one cycle of phase difference embraces 18 green, 24 red and 30 purple lanes. An additional phase difference meter responds to the coarse pattern and gives periodic readings which indicate in turn the correct lane of each pattern within a known zone. Now, instead of actuating a meter, the lane identification transmissions automatically resolve the cycle ambiguity in the individual signals, thus eliminating the lane ambiguity at its source.

Use of Decca

For ease of explanation of the principle of operation, the original equipment of decometers and LI meter have been described. With the decision not to discontinue Decca in the 1990s (as had been a possibility) but to extend its operation until at least 2014, together with the advent of airborne computers, the Decca Navigator became one of the options available on the CDU of say the RACAL RNAV system, and is selected by the HYP (hyperbolic) switch.

When the RNAV computer is fed an initial fix, it starts a search through its data bank for the appropriate Decca chain. It determines the 'best' chain on the basis of nearness to the master station and the hyperbolic pattern geometry. Generally the computer uses two master/slave position lines to determine position, only exceptionally (when close to a master station) using the third position line. When another Decca chain would provide better service, the computer will automatically switch to this new chain.

Pre-flight, after switching on the power supply to the RNAV system and a few seconds' warm-up, the self-test procedure can be completed on the CDU. On the ground and in-flight, by selecting NAVAID on the Display Mode Select switch and HYP on the Sensor Select switch, various Decca presentations can be obtained. The CDU can display particulars of the chain automatically selected and the station in use. If a chain has been selected manually, the computer can be triggered to see if it recommends a better chain. The CDU can be selected to indicate on the alpha-numeric display either the red, green or purple zone, lane and fraction or the derived latitude and longitude, waypoint bearing and distance, etc. Other EFIS present the Decca-derived position on a moving-map display.

Range

300 nm by day; 200 nm by night.

Accuracy

Accuracy of the equipment is very high indeed. If the phase difference is read to the accuracy of 6°, the theoretical accuracy for the red slave would be –

$$\begin{aligned}\text{Lane width} &= 240 \text{ lanes in } 60\,\text{nm}\\ &= \frac{60}{240}\\ &= \tfrac{1}{4}\,\text{nm}\\ \therefore\ \text{Accuracy} &= \frac{6}{360} \times 440\,\text{yds}\\ &= 7\tfrac{1}{3}\,\text{yds}\end{aligned}$$

In practice the accuracy is 1 nm on 95% of occasions at maximum range. The degree of accuracy also depends on the area the aircraft is in in relation to the master and slave stations. See Fig. 18.9.

Errors

(1) ***Height error.*** All Decca charts are made up for ground level propagation. Therefore, a slight error occurs, particularly at very high altitudes.

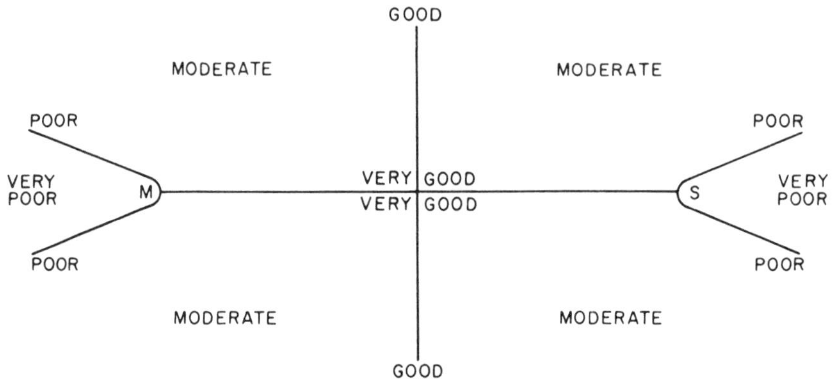

Fig. 18.9 Accuracy spread.

There will be no error, however, when the aircraft is along the perpendicular bisector of the base line, and there will be maximum error when overhead the master or slave station. If high accuracy is required correction charts must be used in conjunction with the readings obtained.

(2) ***Night error.*** Decca assumes that transmissions to the receiver are the shortest distance and the most direct, the ground wave. Being a low frequency aid there will be sky waves present in the aerial at night, when at distant ranges. At night, therefore one must use caution when 200 nm or so from the master. Dusk and dawn are critical periods.

(3) ***Lane slip.*** Although LI signals occupy but half a second, the data flow in the receiver is interrupted for about one second. However, modern equipments include a memory device which takes over when the signals are interrupted to ensure continuous position indication to the pilot.

(4) ***Static.*** Rain static and atmospherics can blot out signals entirely or give incorrect readings through interference.

Decca problems

(1) Given $f = 14.2$, what is the zone width of a purple zone along the base line?

$$\text{Zone width} = \frac{\text{speed of radio wave}}{f} \times \frac{1}{2}$$

$$= \frac{300\,000\,000}{14\,200} \times \frac{1}{2}$$

$$= 10\,560\,\text{m}$$

(2) If $f = 14.26$, what is the width of a purple lane?

$$\text{Purple comparison frequency} = 30f$$

$$= 30 \times 14.26\,\text{kHz}$$

$$= 427.8\,\text{kHz}$$

$$\text{Half wavelength} = \frac{300\,000}{427.8} \times \frac{1}{2}$$
$$= 350.6\,\text{m}$$

(3) Given $f = 14.2$, and there are 14 red zones between the master and the slave.
How many kilometres apart are the two stations?

$$\text{Red zone width} = \frac{300\,000}{14.2} \times \frac{1}{2}$$
$$= 10\,560\,\text{m}$$

For 14 such zones,

$$\text{Width of 14 red zones} = 10\,560 \times 14$$
$$= 147\,840\,\text{m or } 147.840\,\text{km}$$

(4) If the width of a Decca lane along the base line is 440 m, what is the comparison frequency?
Lane width = 440 m; therefore wavelength = 880 m and

$$\text{comparison frequency} = \frac{300\,000}{880}\,\text{kHz}$$
$$= 340.9\,\text{kHz}$$

Chapter 19
OMEGA

VLF is the lowest frequency band in the radio spectrum and comprises the frequencies ranging from 3 kHz to 30 kHz. In this band, the signals suffer the least surface attenuation, and travelling between the surface and the ionosphere by the ducting process, they produce very large ranges. Given sufficient transmission power, ranges of the order of 6000 to 10 000 nm may be obtained.

Thus, VLF is a very attractive frequency band in which to develop long range navigation systems but it was totally neglected until around the mid-1950s. Interest was aroused following Decca's success. The investigations which followed revealed that such a proposition was feasible and that a global navigation system could be produced in the VLF band. Against the advantages of the ranges, there are two disadvantages inherent in any system working with such a low frequency band: its variable propagation characteristics must be thoroughly understood and such a system calls for colossal ground installations.

Omega system

Ground installations. The system consists of eight ground transmitting stations, their locations round the world carefully chosen to give a global coverage, owned and operated by the countries in which they are located, but are monitored (for frequency accuracy and stability and for signal transmission characteristics) by the US Coast Guard. The frequencies used are 10.2 kHz, 11.33 kHz and 13.6 kHz. Of these, 10.2 kHz is the navigation frequency and the other two are transmitted to provide lane identification. Of the eight Omega stations, only three stations are on the air at any one time. The transmission is in the form of pulses of 0.9 to 1.2 second duration and there is a silent period of 0.2 second between the end of one set of three transmissions and the beginning of the next one. The whole non-repetitive sequence takes 10 seconds. For interest, the sequence of transmission from the eight stations is shown in the following table.

Norway	X	Z	Y					
Liberia		X	Z	Y				
Hawaii			X	Z	Y			
N Dakota				X	Z	Y		
La Réunion					X	Z	Y	
Argentina						X	Z	Y
Australia	Y						X	Z
Japan	Z	Y						X

X: 10.2 kHz Y: 11.33 kHz Z: 13.6 kHz

Thus, at the commencement of the sequence, Norway, Australia and Japan transmit together on frequencies, 10.2 kHz, 11.33 kHz and 13.6 kHz respectively. There is a 0.2 second silence and then Norway, Liberia and Japan will transmit, and so on.

Principle of operation. This is a hyperbolic system and the principle has certain similarities with the Decca system. An Omega receiver, synchronised to 10.2 kHz transmission wavelength measures the phase of the signal being received from the selected station. This information is retained in the memory until the phase of a second signal from a different station is received. The difference between the phases of the two stations gives a hyperbolic position line. With a phase measurement from a third station a three position line fix is plotted as follows:

Phase difference between stations A and B – one position line
Phase difference between stations B and C – second position line, and
Phase difference between stations A and C – third position line.

From the above combinations it will be seen that a fourth station will give a six-position line fix.

Airborne equipment. The airborne equipment consists of a small, flush-mounted aerial, a receiver/computer and a CDU. At the commencement of the flight, enter the date, time, present position and the way points in the CDU. The date and time are necessary to make allowances for the variations in the propagation characteristics and the system will accept the position accurate to within 36 nm. Subsequently the present position will be continually indicated as left and right display in latitude and longitude. Depending on the type of the equipment, other types of display may be selected as follows:

XTK/TKE – Cross track error, track angle error
DIS/ETE – Distance to way point, ETA to way point
WV
GS/DA – Ground speed and drift
TK/DTK – Track angle for desired track
CH/OFS – Compass heading, magnetic variation.

The system can operate with magnetic, true or grid compass reference, can hold variation in memory and if less than three stations are being received it will automatically go into DR mode. A typical equipment may accept up to nine way points. The signals may be fed direct to the auto-pilot and the system may be integrated with Doppler or INS. With suitable airborne equipment, Omega signals can be used to provide range position lines, i.e. Omega Rho/Rho. (See later in chapter.)

Lane identification. Just as in the Decca, a lane slip may occur if the signals were temporarily interrupted and a lane ident will be necessary. To enable the lane identification to be carried out, the receiver is supplied with two identification signals, 11.33 kHz and 13.6 kHz. By taking a beat frequency of either of these frequencies with the navigation frequency of 10.2 kHz, the lane may be widened as follows:

10.2 kHz and 13.6 kHz: beat note = 3.4 kHz and

$$\text{lane width} = \frac{300\,000}{3.4 \times 2}\,\text{m} = 44\,\text{km or } 24\,\text{nm (approx)}$$

10.2 kHz and 11.33 kHz: beat note = 1.13 kHz and

$$\text{land width} = \frac{300\,000}{1.13 \times 2}\,\text{m} = 133\,\text{km or } 72\,\text{nm (approx)}$$

Thus, the system is capable of identifying your lane up to 72 nm and the computer will take care of it.

Errors of the system. The errors of the system mainly arise from two sources: propagation anomaly and ground conductivity. Both these factors affect the phases of the signals. Propagation anomaly arises from the effect of the sun's radiation on the ionosphere, which causes the ionosphere level to rise and fall with time. This up/down movement of the ionosphere level affects the radiation pattern of the signals contained under the base of the ionosphere and causes variations in the phase velocity.

Ground conductivity which varies widely over the earth's surface causes varying attenuation of the signals and changes the speed of the radio waves. This has a similar effect on the signal phases. Generally the greatest attenuation occurs over the ice-covered surfaces of Antarctica and the Aretic and the least attenuation over exposed ocean areas.

Both these errors have been thoroughly investigated, the variations with time and place are known and are allowed for in the computer. Any resultant inaccuracy of the system arises from incorrect predictions of the behaviour of the signal phases. There is a programme for keeping the propagation characteristics under observation and updating the present information as new data become available.

Apart from the above, there are frequent unpredicted variations in the ionospheric state on account of the sudden flare-up of sun spots, the disturbance reaching us as X radiation. These may simply cause lane slips or at worst, a total disruption of the signals. Even here, however, a limited warning is possible in advance. The warnings of any abnormal conditions are given out in the NOTAMS and in some countries there are regular broadcasts as well. The station status and the warnings of any unpredicted conditions are checked by the crew as a matter of pre-flight briefing. The overall accuracy of the system is given as 1 nm during the day, 2 nm at night.

Another VLF system is that known as Rho/Rho or Ranging system, or just VLF system.

Ranging system (Rho-rho system)

Instead of transmitting its own signals, this system makes use of the continuous communication signals transmitted by the US Navy, with the Navy's agreement, of course. Its transmitters are suitably located round the world and transmitting in the frequency range of 15 kHz to 30 kHz provide a worldwide coverage.

The principle of the system is that the signals from different stations arrive at a given position at different times. If the time taken by the signals to travel from the transmission source to a given point can be measured, this time can be converted to the distance from the transmitter and plotted as a circular range position line (rho). In favourable circumstances, signals from any two conveniently placed stations would give a navigational fix (R/R). In actual practice, because of the high transmission power (from 50 to 1000 kW) and the use of very low frequencies it is quite common to receive more than two stations anywhere at any given time.

The principle is implemented by use of an accurate on-board atomic time standard which produces a timing pattern identical to the received signals. The necessary navigational data are derived from the comparison of the phase of the incoming signal with the time standard, one station giving one position line.

Several receivers capable of utilising these VLF signals are now on the market: a typical system would operate as follows. The airborne equipment consists of a flush-mounted aerial, receiver/computer, an atomic time standard, control head with all its sixteen buttons on the keyboard, a display head and a station indicator.

On switching on the equipment the station indicator lights up to indicate the stations being received. The receiver automatically selects the best stations for its operation. After around five minutes of warming up time from the initial power supply, the atomic standard will have stabilised and a light signal will indicate that the system is ready to navigate. The computer is initially fed with the accurate start position and a continuous display of the present position is then maintained. Other types of display may be selected: distance to go, distance across, heading to a way point, ground speed, time to a way point and so forth. As the flight progresses the receiver will continue to adjust to the best incoming signals. Basically a very reliable system, it may lose signals when flying in cloud with high ice crystal content. Then the system will go in the DR mode but it will pick up from where it left off when stronger signals are received again.

Test questions

(1) The correct name for the 'Omega' fixing system is:
(a) Omega (b) Oceanic Omega (c) Omega Radar System.

(2) Omega is:
(a) a Doppler navigation system (b) a long range radar system
(c) a hyperbolic navigation system.

(3) Omega operates on:
(a) 15 kHz, 30 kHz, 45 kHz (b) 10.2 kHz, 11.33 kHz, 13.6 kHz
(c) 15.2 kHz, 21.33 kHz, 30.6 kHz.

(4) Omega operates in the frequency band:
(a) VLF (b) VHF (c) L-band.

(5) In the Rho/Rho ranging system, the Rho indicates:
(a) a range position line (b) a radar position line
(c) a radio bearing position line.

Chapter 20
SATELLITE-ASSISTED NAVIGATION (GNSS/GPS)

Like many other radio navigation aids, satellite assisted navigation systems were initially developed for military use but the facilities on which they are based are now made available to civil users. Global navigation satellite systems (GNSS) have become possible by the launching of constellations of satellites orbiting the earth, each satellite carrying an extremely accurate atomic clock. Basically, by timing the period that is taken for a satellite's transmission from a known position to reach the aircraft's receiver, the distance satellite to receiver is determined so enabling a range position line, or strictly a section of the surface of a position sphere with centre the satellite, to be established. Signals received simultaneously from three or more satellites enable a fix to be obtained and simultaneous ranges from four satellites provide a three-dimensional fix (Fig. 20.1). GNSS can therefore be

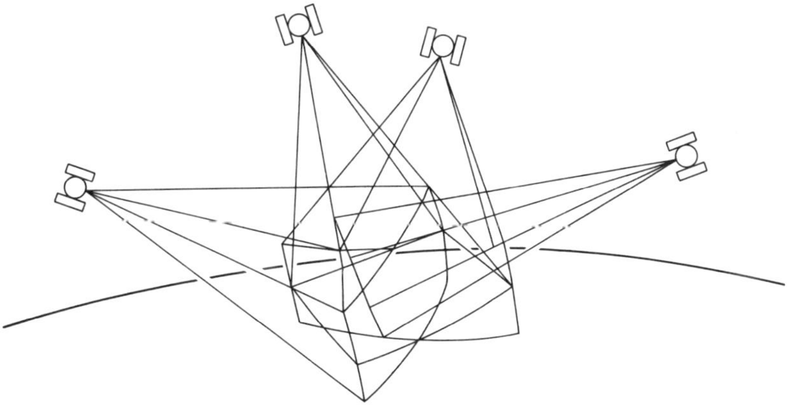

Fig. 20.1 Fix by GNSS.

claimed to be a navigational system independent of ground facilities and when fully developed it will also serve as an approach aid, superseding ILS and MLS in the 21st century. However more of this later at the end of this chapter.

Satellite constellations

The global positioning system (GPS) constellation of satellites is a military facility under the auspices of the US Department of Defense (DoD). It will

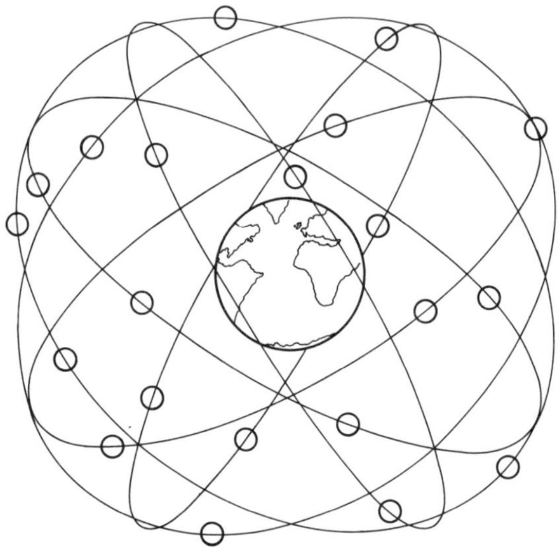

Fig. 20.2 GPS constellation.

have 21 active satellites plus three operational spares in six circular orbital planes around the earth (Fig. 20.2) inclined at an angle of 55° to the equator. Each satellite weighs about one metric tonne, has an expected life of about seven to eight years and orbits the earth once every twelve hours at a height of 20 200 km (10 900 nm). At least four GPS satellites should always be in view from any place in the world enabling highly accurate determination of latitude, longitude, altitude, velocity and time from the satellites' signals received by an aircraft in flight or on the ground.

The Soviet Union, now the CIS, also launched a broadly similar system

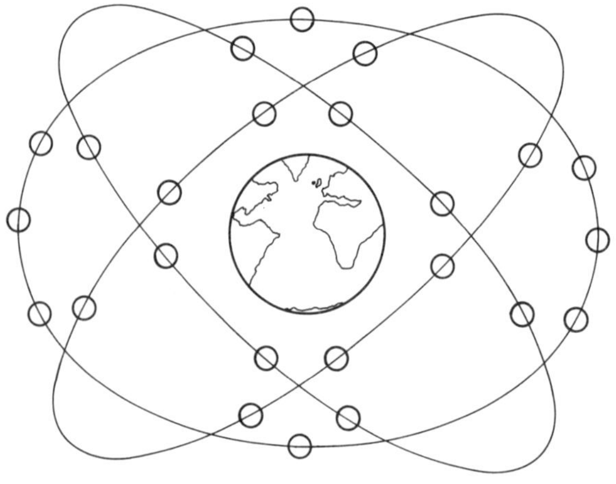

Fig. 20.3 GLONASS constellation.

(global orbiting navigation satellite system) (GLONASS) which too will have 21 active and three spare satellites with the constellation in this case in three orbital planes inclined at an angle of 60° to the equator (Fig. 20.3).

As with GPS, although originally provided as a military facility, GLONASS is also available for civil aviation use, the satellites having a 12-hour orbit at a height of 19 000 km (10 250 nm).

Fixing position

The satellites used in GNSS/GPS give the same information as a DME, i.e. the distance of the aircraft from the beacon. Consider a DME giving a range of 200 nm. This locates the aircraft on the surface of a sphere centred on the beacon and with a radius of 200 nm. Normally we assume that the aircraft is on the surface of the earth, as even at FL 350 this only represents a negligible error and therefore we can locate the aircraft on a small circle where the two position spheres (of the earth and of the DME range) meet.

To obtain a fix a second DME range may be used and now the aircraft will be located at one of two positions where the two DME range spheres cut each other and the surface of the earth. To resolve this ambiguity requires a third DME to be used. The intersection of the three DME range spheres and the surface of the earth will give a positive fix. Notice that we are using four position spheres (the earth and three DMEs). If no assumption is made about the aircraft's location in relation to the earth, a fourth range sphere will be required. This will then provide a fix in space – latitude, longitude and altitude, (i.e. a 3D fix). Four satellites are therefore required for 3D navigation. In GPS, if only three satellites are available, the aircraft altitude can be fed in manually so effectively producing a fourth position sphere with its centre at the centre of the earth and its radius the earth's radius plus aircraft altitude.

GPS satellites transmit on two L-band frequencies of 1575.42 MHz (L_1) and 1227.6 MHz (L_2) although the satellites' control is through up-link and down-link frequencies (2227.5 MHz and 1783.74 MHz respectively) in the S-band. The L_1 signal is modulated with two pseudo-random codes: P (precise) code and C/A (coarse acquisition) code, while the L_2 signal is modulated with the P-code only. Only the L_1 C/A code is available to civil aircraft and although having an inherent accuracy of fixing position within about 30 m, it is deliberately degraded to an accuracy of approximately of 100 m (on a 95% probability basis) so that no-one can use the system effectively, for example for the guidance of so-called 'smart bombs', contrary to the United States' national security. (For military aircraft using also the L_2 P-code, the claimed accuracy is 1 to 3 m.) The 'pseudo-random' coding of the signals is so called because although it may appear to be random, in fact it is generated according to a complicated set of instructions, repeating itself every seven days. The whole C/A code has a duration of one millisecond.

Both the satellite and the GPS receiver have to generate exactly the same radio signal at exactly the same time and to embody the identical coding to

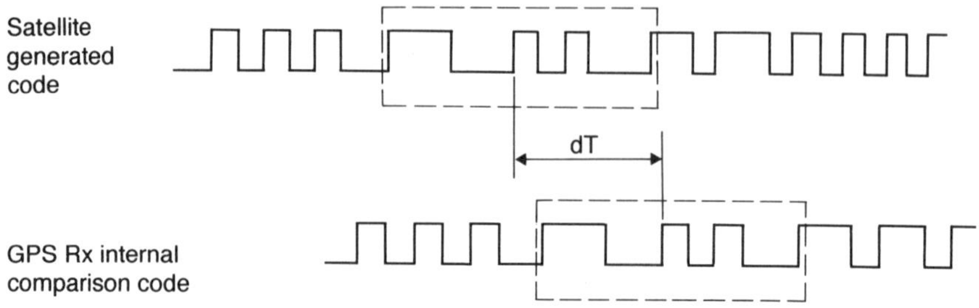

Fig. 20.4 Measurement of dT.

enable the signals to be compared (Fig. 20.4) although the GPS comparison signal is not transmitted externally. As shown diagrammatically in Fig. 20.4, the GPS receiver matches up the code received from the satellite with the code that the GPS receiver has been generating for its own internal timing system to determine the time interval (dT) since it was generated.

If for example, dT is 0.1 second, then the signal was transmitted from the satellite 0.1 seconds earlier when it was

$$0.1 \text{ seconds} \times 186\,000 \text{ miles/sec} = 18\,600 \text{ miles}$$

from the aircraft.

The satellite's transmission also includes the precise time of the transmission, its orbital position, atmospheric propagation data and any satellite 'clock bias' information. While the satellites' timepieces are atomic clocks and the GPS receiver clocks are accurate crystal oscillators there is nevertheless a timing bias error or clock bias in the measurement of the time

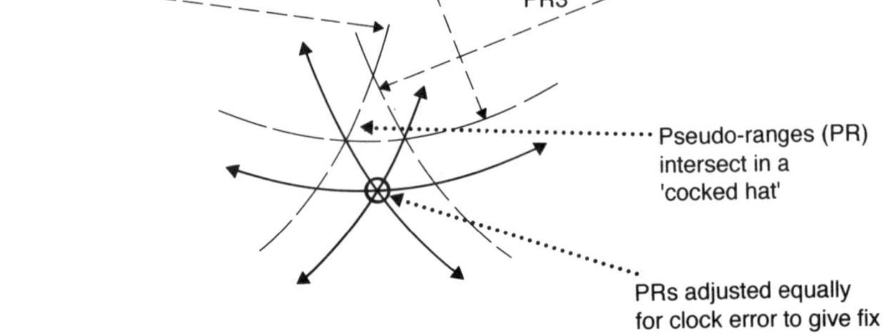

Fig. 20.5 Removal of clock bias.

interval and hence the range, so the measured ranges are called pseudo-ranges. However as the clock bias at the receiver will be the same for all of the pseudo-ranges measured at a particular time it can be removed by using three simultaneous equations (one for each pseudo-range) to solve for the three unknowns of clock error, latitude and longitude, as illustrated in Fig. 20.5.

Other errors inherent in GNSS are

Satellite clock error. The integrity of the satellites is continuously monitored and if the satellite clock becomes even a few nano-seconds in error (1 nano-second = 0.000 000 001 second) it is corrected so that the stated error due to this cause is up to 0.5 m in range.

Satellite ephemeris error. This occurs if the satellite is not precisely where it is believed to be in its orbit. The satellite control however is so exacting that the range error due to this cause is also quoted as kept within ±0.5 m – i.e. the width of the pilot's panel!

Atmospheric propagation error (ionosphere error). The density of the charged particles (ions) in the atmosphere changes diurnally and seasonally. The delay due to the effect of the ionosphere on the satellite signal also depends upon its angular path to the receiver but fortunately this error is predictable and not expected to cause a range error of more than 4 m.

Instrument/receiver error. This may arise from electrical noise, errors in matching the pseudo-random signals and computational errors but even then only to the order of a range error of one metre.

Satellite geometry error. This is not truly a GPS error at all but a basic principle when using position lines to obtain a fix. If the angle of cut of two position lines is very shallow, the effect of any error is much greater than if the position lines intersect at a right angle or nearly so (Fig. 20.6). In the

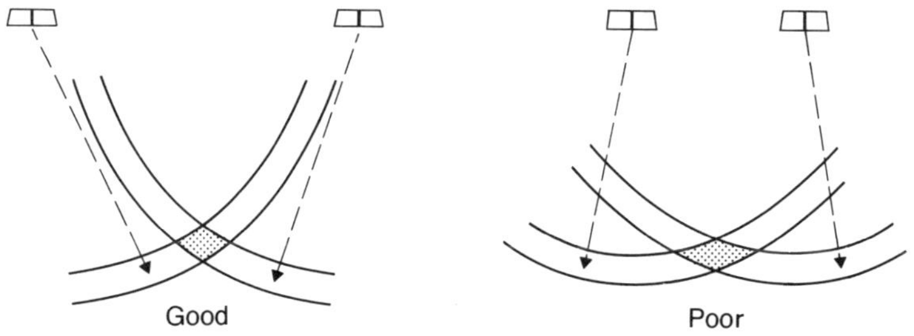

Fig. 20.6 Satellite geometry error.

application of GNSS, this means that if two or more satellites are close together when being used for two-dimensional navigation, range position lines obtained from them will intersect at a shallow angle and any signal error effect is therefore larger. The term used in GPS is *'geometric dilution of precision'* (*GDOP*) which may be sub-divided into time dilution of precision – the range equivalent of clock bias (TDOP) and position dilution of precision – based on satellite geometry less clock bias (PDOP), which in turn may be sub-divided into vertical and horizontal components (VDOP and HDOP). It is in order to avoid this source of error that the GPS constellation has its orbital planes all inclined at 55° to the equator; each orbital plane has four satellites; each orbital satellite takes 12 hours to complete an orbit. GNSS receivers which can use both GPS and GLONASS satellite data simultaneously will suffer least from the problem of satellite geometry.

GPS differential correction (dGPS)

The only significant errors in fixing position by GPS are atmospheric (ionosphere) error and the deliberate error introduced by the DoD, called selective availability (SA) which as already mentioned is designed to prevent unfriendly users exploiting the extraordinarily high accuracy of GPS navigation to threaten the US with offensive military actions. For civil aviation a correction factor – called the differential correction – can be determined for aircraft navigation which eliminates the effects of ionospheric error and SA, which is then transmitted to the aircraft (by ACARS or similar – see chapter 3). The principle is shown diagrammatically in Fig. 20.7. GPS signals are

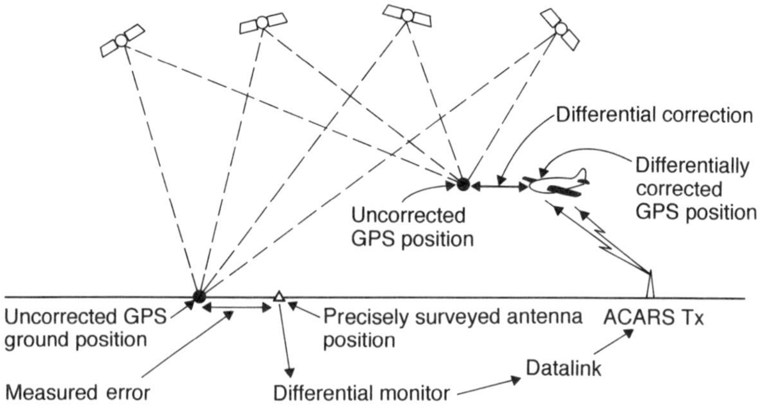

Fig. 20.7 Differential GPS.

received at a station of which the antenna position has been precisely surveyed. The monitoring ground station computes its position based on the incoming satellite signals. The difference between the surveyed (true) position and the computed (raw) position is the differential correction – a vector with direction and magnitude in three dimensions. When this correction is applied to the raw position the refined position will be accurate to 1 to 3 m. The dGPS correction is sent by ACARS uplink to the aircraft's GPS sensor if it is within 70 nm of the dGPS station. The dGPS correction is updated periodically and sent via the datalink to GPS-navigating aircraft.

Airborne equipment

Externally there is a small streamlined antenna which is normally mounted on the fuselage, while on the flight deck the receiver/CDU appears in various models with either a CRT or more likely a LCD presentation giving various alternative items of information such as:

- Latitude, longitude, altitude
- Offset range and bearing from a specified waypoint (WPT)
- Parallel track navigation
- TAS, GS, TMG, desired track
- Heading to steer and time to WPT
- Time, upper winds, magnetic variation
- Alerts when approaching pre-set positions, areas, etc.
- Self-test facility
- Range, distance, bearing
- Vertical navigation
- Estimated en-route time, ETA
- Satellite status
- Fail/warning signals

The controls range from push/push or rocker type switches for ON/OFF, BRIGHTNESS, etc., to an ALPHANUMERIC keyboard to enable pilots to input/extract any of the listed information. The GNSS unit can also be interfaced with EFIS, autopilot and moving-map displays. For the pilots of general aviation (GA) aircraft, personal portable models are available which can be unplugged from the panel and the aircraft's power supply, to be taken to the office, club or home to plan the next flight or to the marina for use on a yacht using portable batteries.

Advantages/disadvantages

Advantages:

(1) Capable of use worldwide at all operating speeds and altitudes
(2) Highly accurate
(3) Available throughout the 24 hours with no diminution of accuracy
(4) Provides three-dimensional positioning on a continuous basis
(5) Unrestricted range
(6) Entirely space-based
(7) Coupled with inertial navigation systems, can make terrestrial-based aids such as NDBs, VORs, DMEs, etc., redundant
(8) When GPS/GLONASS development is complete and all satellites available, the accuracy of GNSS will enable aircraft separations to be reduced

(9) Similarly to (8), GNSS when perfected will be able to supplant MLS and ILS as a precision approach aid
(10) For light aircraft, there are lightweight, relatively inexpensive and portable models available.

Disadvantages:

(1) Basically it is a military system which civil aircraft are allowed to use; currently the US has agreed to providing GPS to civil users free of charge until AD 2002 and similarly the CIS is providing GLONASS free until AD 2007
(2) As a military system, the existing GNSS may conceivably be interrupted if a serious military need develops
(3) The system, although it has evolved rapidly, is still technically under development
(4) Not as accurate in the civil mode as it is in the military mode (by deliberate policy)
(5) Present system cannot guarantee the pilot with immediate indication of satellite malfunction or GNSS degradation; this makes it currently unsuitable as an approach aid for high category landings.

Latest developments

Chapter 8 referred to international disagreement on the use of MLS and DGPS as a future landing aid. Europe favours the continued use of MLS until DGPS has the same CAT3 performance ability. America wishes to phase out MLS and concentrate on the development of DGPS to the same CAT3 capabilities and integrity. This they hope to achieve much sooner (2001) than Europe deem possible (2015)!

Undoubtedly DGPS will provide the final answer. Meanwhile ICAO is approving both systems and a multi-mode receiver (MMR) for their reception.

Test questions

(1) Position fixing by satellite is by means of:
  (a) simultaneous bearings from a number of satellites
  (b) range and bearing from each known satellite's position
  (c) simultaneous ranges from a number of satellites.

(2) The dGPS is:
  (a) the time difference between the transmission of the satellite's coded signal and the time that it is received at the aircraft
  (b) the time difference between the satellite and the aircraft GPS unit transmitting the same code point on the signal
  (c) the correction applied to the 'raw' GPS position to obtain a more precise position.

(3) The reason that the satellites of GPS and GLONASS do not collide is:
   (a) the GPS orbit is at 55° to the equator while the GLONASS is at 60°
   (b) the GPS satellite orbits are 1200 km higher than the GLONASS orbits
   (c) GLONASS has only three orbital planes while GPS has six orbital planes.

(4) The GPS can:
   (a) provide two-dimension and three-dimension positions
   (b) only provide position in latitude and longitude
   (c) indicate two-dimension positions or position lines in the horizontal plane.

(5) The GPS transmission frequency which can be used by civil aircraft is:
   (a) 1575.42 MHz in the L-band
   (b) 2227.5 MHz in the S-band
   (c) 4454.5 MHz in the C-band.

(6) The US DoD degrades GPS accuracy:
   (a) to protect US national security
   (b) to protect the DoD copyright
   (c) to prevent 100% reliance on a system which may have to be interrupted for military purposes at a moment's notice.

(7) The accuracy of range measurement of satellite to GPS receiver is:
   (a) seriously reduced during periods of sunspot activity
   (b) seriously eroded during periods of anomolous propagation in an inversion
   (c) derived from accurate timepieces in the satellite and the aircraft's GPS receiver.

(8) Satellites used in GNSS, orbit the earth:
   (a) once in twelve hours (b) once in eight hours
   (c) once in 24 hours.

(9) Satellite geometry error is greatest when:
   (a) satellites are closest together
   (b) satellites are spaced well apart
   (c) satellites are nearest the horizon.

(10) It may be claimed that GNSS has the advantage over other air navigation systems in that it:
   (a) is basically a military-based system made available to civil users with solar-powered satellites which have an infinite life
   (b) has no possible input by the pilot
   (c) is entirely space-based.

Chapter 21
TRAFFIC COLLISION AVOIDANCE SYSTEM (TCAS)

In the earlier chapter on SSR reference was made to the enhanced detail of data from aircraft (both actual and flight intentions) via Mode S, hopefully leading to better information on conflicting traffic and resolution of potential problems together with unambiguously identified aircraft. ICAO Annex 10 which specifies the SSR standards and procedures also covers ACAS (Airborne Collision Avoidance System). It is an unfortunate fact that mid-air collisions although rare, have occurred since the first one in 1910 in which the Wright brothers were involved. The investigators' collision reports over the years, have concluded that the various accidents have been attributable to a number of causes.

Particularly since the collision over Grand Canyon in the mid-1950s, strenuous efforts have been and are being made to devise a system to avoid any future collisions and the equipment development has been accorded a high priority. ICAO defines ACAS at three levels:

(1) An independent ACAS (I) which provides traffic advisories (TAs) as an aid to initiate see-and-avoid action but does not include the capability for generating resolution advisories (RAs)
(2) An independent ACAS (II) which provides vertical resolution advisories (RAs) in addition to traffic advisories
(3) An independent ACAS (III) which provides both vertical and horizontal RAs as well as TAs.

Arising from these the design of ACAS is on the lines of:

(1) the aircraft carries not only a transponder but also a means to interrogate other aircraft transponders, making air-to-air interrogation and reply possible. (It is an interesting fact that over 50 years ago in the Second World War, allied aircraft carried IFF (identification friend or foe))
(2) replies from aircraft up to a range of 30 nm are triggered
(3) the reply data is used in the airborne computer to assess the threat if any, from any nearby aircraft
(4) the pilot has a display of the approximate relative position of nearby aircraft together with their altitude if the other aircraft are fitted with altitude decoders
(5) the reply data is put into one of two categories:
    (a) those aircraft which constitute no threat, resulting in a TA
    (b) those aircraft which do present a threat, resulting in an RA being given to the pilot advising him to manoeuvre to avoid collision.

The US Congress has pressed the Federal Aviation Administration (FAA) for the introduction of ACAS and legislation requiring aircraft built to carry more than 30 passengers to have an approved system for flying in US airspace. Although ICAO named the system ACAS, in fact it is usually known as traffic collision avoidance system or as traffic alert and collision avoidance system. Because it is an entirely airborne decision and action system, it is fair to say that air traffic controllers have had reservations over its introduction, which in Europe is following a more cautionary introductory programme than in the US.

TCAS I, corresponding to ACAS I, is the simplest system and intended just to enable pilot to visually acquire nearby traffic. It will detect and display range and approximate relative bearing. If the other (target) aircraft is carrying Mode C or Mode S SSR it will also display relative altitude. The system will alert pilots visually and aurally – TRAFFIC TRAFFIC – if the conflicting traffic comes within 40 seconds of a potential collision, i.e. a TA. It does not indicate a course of action to follow to avoid the target aircraft. Currently the FAA is requiring all smaller aircraft with 30 or fewer seats (i.e. commuter or feeder airliners) to carry TCAS I in US airspace by 1995.

TCAS II, corresponding to ACAS II, now being required in US airspace as already mentioned for larger airliners, has greater range and bearing accuracies than TCAS I. It will offer avoidance messages as Resolution Advisories, CLIMB, INCREASE CLIMB, DESCEND, INCREASE DESCENT. If your aircraft and the target aircraft both have Mode S datalink transponders, the TCAS II computers on board the two aircraft will coordinate their RAs to offer complementary vertical avoidance manoeuvres to each flight crew. The TCAS II airborne system comprises a radar transponder, directional antennae on the top and bottom of the fuselage, computer processor and software with a flight-deck CDU.

The visual advisory may take one of two general forms, varying on the equipment kit determined by the operator. As the avoidance manoeuvre is simply to climb or descend, the simplest instrumentation is in the form of a modified conventional vertical speed indicator (VSI) known as a RA/VSI. When an RA is generated, on one side of the 0 vertical speed position a green arc appears and on the other side of the 0 rate of climb/descent position a red arc appears around the edge of the VSI dial to indicate safe (green) and dangerous (red) vertical manoeuvre to avoid the potential collision.

The second type of presentation is by using a radar display. This may be on a modified airborne weather radar, on a dedicated TCAS traffic display CRT or if EFIS is fitted, (see *Ground Studies for Pilots volume 3*) on the EFIS navigation display (ND). The convention on airborne radar traffic displays is that, shown on the approximate bearing and range from one's own aircraft, are:

'Other' traffic not offering a threat: a hollow white or blue (cyan) diamond outline

'Proximate' traffic within 6 nm and ±1200 ft: a solid white or solid blue (cyan) diamond
Traffic advisory (TA): a solid yellow circle
Resolution Advisory (RA): a solid red square

Each of these aircraft symbols is accompanied by a 'data tag' the same colour as the symbol. The data shown comprises a number to show the other aircraft's height relative to one's own in hundreds of feet, i.e. +02 indicating 200 ft above. If the target is climbing or descending (at more than 500 ft/min) an up or down arrow beside the number indicates this. If the target is not reporting altitude, although the radar will show the aircraft's relative position, there will be no data tag. On those models of EFIS where the primary flight display (PFD) embodies a strip type of VSI along one side, when an RA is generated a corresponding indication may be displayed on the strip. Part will be illuminated green as the 'fly to' area while part of the strip is illuminated red as the forbidden manoeuvre.

TCAS III, corresponding to ACAS III, at the time of publication is still being developed. When it is operational, as originally defined by ICAO, it will offer manoeuvres in the vertical and horizontal planes. This will require the target aircraft's path to be tracked accurately. In TCAS II, measurement and presentation of the relative bearing is sufficient to indicate the risk of conflict but with TCAS III, it must be known precisely and it is particularly this element which is causing the longer development time.

Reservations and limitations

(1) The concern of ATCOs over the introduction of TCAS has been mentioned earlier. The traditional principle may be quoted of 'Navigation in the air – Control from the ground' and the possible conflict between ATC instructions issued to a pilot and what his TCAS is advising. For example, in a terminal control area (TMA) in an aircraft fitted with TCAS, before the carriage of TCAS is mandatory in that TMA. The ATC units throughout Europe, especially those newly built, will have equipment compatible with or technically in advance of TCAS by means of which pilots' clearances and routeings will be monitored. Pilots should not abandon their legal separation requirements. An ICAO Circular of 1985 warned "Manoeuvres initiated as a result of ACAS indications must not result in unsafe reduction of separation between affected aircraft and other non-conflicting aircraft in the same air traffic environment". A possible future development has been suggested, as well as TCAS III, of a G-CAS for airspaces such as TMAs. This would be a ground-based collision warning system when Mode S equipments are commonplace, bringing controllers firmly back into the conflict-resolution loop.

(2) It is to be hoped that by the time that TCAS is a legal requirement in Europe, experience in the USA will have eliminated, or action taken to combat, false alarms which accompanied TCAS introduction. The

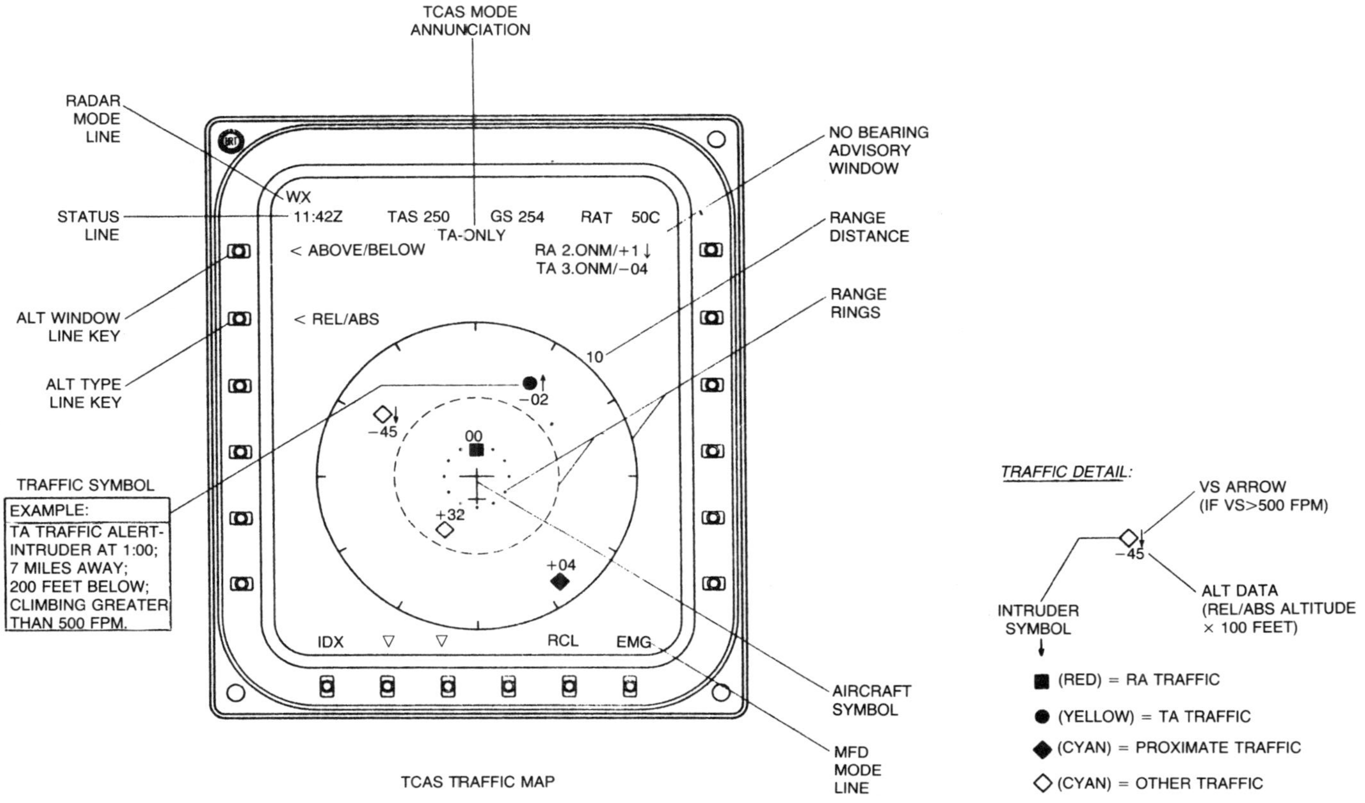

Fig. 21.1 TCAS traffic map.

phantom RAs, called parrots, arose from a variety of sources – some very basic. Surface transponders on ships, ATC test units and parked aircraft for example returned parrots. Ships' returns can be eliminated by making them report nil altitude. Another cause of parrots was leakage internally between the aircraft's transponder signal and the TCAS so that in effect the pilot received a false resolution advisory to avoid his own aircraft!

(3) Several situations will require the inhibition of TCAS, totally or partially, to avoid conflicts with other operational needs such as:
   (a) when the GPWS or windshear alerting system is in operation and should take priority
   (b) below certain altitudes anyway to ensure terrain clearance, such as no aural alerts below 400 ft agl; no RAs below 500 ft agl; no DESCEND RAs below 700 ft agl; no INCREASE DESCENT RAs below 1800 ft
   (c) no CLIMB or INCREASE CLIMB RAs when the aircraft is near its performance limits or above a certain altitude

(4) In some manoeuvring situations such as when parallel approaches are being carried out, the pilot to be able to opt to select a TA only mode, so eliminating distracting RAs.

Test questions

(1) In a TCAS, RA stands for:
   (a) Radar advisory (b) Resolution advisory (c) Radar activate.

(2) On a TCAS, TA stands for:
   (a) Traffic advisory (b) Threat advisory
   (c) Temporary advisory.

(3) The aural warning on TCAS accompanying a TA is:
   (a) THREAT THREAT! (b) TRAFFIC TRAFFIC!
   (c) WHOOP WHOOP!

(4) Outside terminal airspace, the range for TCAS operation is normally:
   (a) 100 km (60 nm) (b) 50 km (30 nm) (c) 20 km (12 nm).

(5) On TCAS II, the aural message is one of:
   (a) CLIMB, DESCEND, INCREASE CLIMB, INCREASE DESCENT
   (b) CLIMB, DESCEND, TURN LEFT, TURN RIGHT
   (c) THREAT THREAT – CLIMB, THREAT THREAT – DESCEND.

(6) On an EFIS display, Proximate traffic is shown as:
   (a) solid red square (b) solid yellow circle
   (c) solid cyan or solid white diamond.

(7) The distance to proximate traffic is within:
   (a) 1200 ft and 6 nm (b) 600 ft and 12 nm (c) 600 ft and 30 nm.

(8) On a RA/VSI, the eyebrow FLY TO vertical speed indication is shown in:
(a) Green (b) Gold (c) Cross-hatched white.

(9) The difference between TCAS I and TCAS II is that:
(a) TCAS II can only be fitted in large aircraft (with more than 30 seats) but TCAS I can be fitted in any size of aircraft
(b) TCAS II can generate TAs and RAs while TCAS I can only generate traffic advisories
(c) TCAS II can only be fitted to aircraft equipped with full EFIS but TCAS I can be fitted to any aircraft with SSR.

(10) A resolution advisory is shown on the ND panel of EFIS as:
(a) a solid yellow circle (b) a solid white diamond
(c) a solid red square.

GLOSSARY OF ABBREVIATIONS

a.c.	alternating current
a/c	aircraft
ACARS	aircraft communications addressing and reporting system
ACAS	airborne collision avoidance system
ACC	Area Control Centre
ADF	automatic direction finding equipment
ADI	attitude direction indicator
AF	audio frequency
AFTN	aeronautical fixed telecommunication network
agl	above ground level
AIC	aeronautical information circular
AIP	Aeronautical information publication
AM	amplitude modulation
amsl	above mean sea level
ANO	Air Navigation Order
ARINC	Aeronautical Radio Inc, a non-profit organisation owned by member operators to define form, fit and function of aviation avionics equipment
ASB	alternating sideband
ASR	approach surveillance radar
ATC	Air Traffic Control
ATCC	Air Traffic Control Centre
ATS	Air Traffic Service
AWR	airborne weather radar
Bcn	beacon
BFO	beat frequency oscillator
brg	bearing
B-RNAV	basic area navigation
CAA	Civil Aviation Authority
CAVOK	weather (in general terms) fine and clear (*cf volume 4*)
CDU	control and display unit
CH	compass heading
CL	centreline
cm	centimetre(s)

CPU	central processing unit
CRT	cathode ray tube
c/s	callsign
CW	carrier wave/continuous wave
DA	drift angle
dB	decibels
d.c.	direct current
DDM	difference in depth of modulation
Dev	deviation
DF	direction finding
dGPS	GPS differential correction
DH	decision height
diff	difference/differential
dir	direction
dist	distance
DME	distance measuring equipment
DOC	designated operational coverage
DoD	Department of Defense (USA)
DR	dead reckoning
dT	time interval
DTK	desired track
duplex	separate channels for transmission and reception
DVOR	Doppler VOR
°*E*	degrees east
EFIS	electronic flight instrument system
EHF	extremely high frequency
ELF	extremely low frequency
ETA	estimated time of arrival
FAA	Federal Aviation Administration
f_d	Doppler shift
FM	frequency modulation
FMS	flight management system
ft	feet
ft/sec	feet per second
GCA	ground controlled approach
GDOP	geometric dilution of precision
GES	ground earth station
GHz	gigahertz
GLONASS	global orbiting navigation satellite system (CIS)
GNM	ground nautical miles
GNSS	global navigation satellite system (GLONASS and GPS)
GP	glidepath

GPS	global positioning system (USA)
GPWS	ground proximity warning system
GS	ground speed/glideslope
Hdg	heading
Hdg(M)	heading magnetic
HF	high frequency
Hz	hertz
ICAO	International Civil Aviation Organisation
IFR	instrument flight rules
ILS	instrument landing system
IM	inner marker
INMARSAT	International Maritime Satellite Organisation
INS	inertial navigation system
ISA	International Standard Atmosphere
kHz	kilohertz
km	kilometre(s)
kt	knot(s)
LCD	liquid crystal display
LCZ	ILS localiser
LED	light emitting diode
LF	low frequency
LI	lane identification
LLZ	localiser
LOC	localiser
LMM	locator middle marker
LOM	locator outer marker
LORAN	long range aid to navigation
L/R	left/right
LUHF	lowest usable high frequency
m	metre(s)
MF	medium frequency
MHz	megahertz
min	minute(s)
MKR	marker
MLS	microwave landing system
MM	middle marker
MN	magnetic north
MP	multipulse
mph	miles per hour
m/sec	metres per second
MU	management unit

MUF	maximum usable frequency
μs	microsecond(s)
NDB	nondirectional radio beacon
nm	nautical mile(s)
OBS	omnibearing selector
OCL	obstacle clearance limit
OM	outer marker
°*P*	degrees port
PAR	precision approach radar
PFD	primary flight display
PPI	plan position indicator
pps	pulses per second
PRF	pulse recurrence frequency
PRI	pulse recurrence interval
P-RNAV	precision area navigation
PRP	pulse recurrence period
PRR	pulse recurrence rate
QDL	series of bearings
QDM	aircraft's magnetic heading to steer in zero wind to reach the station
QDR	magnetic bearing from the station
QE	quadrantal error
QGH	landing procedure clearance
QTE	true bearing from the station
QTF	request position fix
°*R*	degrees relative
RA	radio altitude/resolution advisory (in TCAS)
RAD	approach surveillance radar
RBI	relative bearing indicator
rel	relative
Rel Brg	relative bearing
RF	radio frequency
RMI	radio magnetic indicator
RNAV	area navigation
ROD	rate of descent
RTF	radiotelephony
RVR	runway visual range
RW/RWY	runway
Rx	receiver
°*S*	degrees starboard
SATCOM	satellite communication

SATNAV	satellite navigation
SELCAL	selective calling system
SHF	super high frequency
Sig	signal
SSB	single sideband
SSR	secondary surveillance radar
STAR(s)	standard terminal arrival route(s)
Stb	starboard
STBY	standby
STOL	short take-off and landing
T, (T)	true
TA	traffic advisory (in TCAS)
TAP	terminal approach procedures
TAS	true airspeed
TCAS	traffic collision advoidance system
	traffic conflict alert system
TKE	track error
TMG	track made good
TR	track
TVOR	terminal VOR
Tx	transmitter
UHF	ultra high frequency
UK	the United Kingdom
UKAIP	the United Kingdom aeronautical information publication (Air Pilot)
UTC	co-ordinated universal tune
Var	magnetic variation
VDF	VHF direction finding
VF	voice frequency
VFR	visual flight rules
VHF	very high frequency
VLF	very low frequency
VOR	VHF omnidirectional radio range
VOT	a test VOR
°W	degrees west
WDGS	windshear detection and recovery guidance system
WPT	waypoint
WT	wireless telegraphy
WV	wind velocity
Wx	weather

ANSWERS TO MULTI-CHOICE TEST QUESTIONS

Chapter 1
Q11 (b), Q12 (a), Q13 (a), Q14 (c), Q15 (b).

Chapter 2
Q11 (c), Q12 (b), Q13 (c), Q14 (b), Q15 (c).

Chapter 3
Q3 (b), Q4 (b), Q5 (c), Q6 (b), Q7 (c), Q8 (b), Q9 (a), Q10 (b).

Chapter 4
Q1 (b), Q2 (c), Q3 (b), Q4 (c), Q5 (a).

Chapter 5
Q11 (a), Q12 (a), Q13 (c), Q14 (c), Q15 (a).

Chapter 10
Q4 (b), Q5 (c), Q6 (b), Q7 (c), Q8 (a), Q9 (b), Q10 (a).

Chapter 13
Q3 (b), Q4 (c), Q5 (a), Q6 (a), Q7 (b), Q8 (c), Q9 (a), Q10 (b).

Chapter 14
Q1 (b), Q2 (a), Q3 (b), Q4 (b), Q5 (b).

Chapter 17
Q1 (c), Q2 (b), Q3 (a), Q4 (a), Q5 (b).

Chapter 19
Q1 (a), Q2 (c), Q3 (b), Q4 (a), Q5 (a).

Chapter 20
Q1 (c), Q2 (c), Q3 (b), Q4 (a), Q5 (a), Q6 (a), Q7 (c), Q8 (a), Q9 (a), Q10 (c).

Chapter 21
Q1 (b), Q2 (a), Q3 (b), Q4 (b), Q5 (a), Q6 (c), Q7 (a), Q8 (a), Q9 (b), Q10 (c).

Index

abbreviations, 236–40
ADF (*see* automatic direction finding)
airborne collision avoidance system (ACAS), 230
airborne weather radar (AWR), 156–68
aircraft communications addressing and reporting system (ACARS), 40
altimeter, radio, 169–73
amplitude, 2
amplitude modulation, 9
angle of lead, 63
anode system, 128
anomalous propagation, 29
area navigation, 141
attenuation, 19, 32
automatic altitude telemetering, 146
automatic direction finding (ADF), 48–71
  range and accuracy, 64–67
  use of, 56

beacon saturation, 139
bearings, classification of, 45
beat frequency oscillator (BFO), 55
beat note, 55
BFO (*see* beat frequency oscillator)

cardioid, 50
categories (ILS), 112
cathode ray tube, 127
clock bias, 224
cloud warning radar, 156–68
coastal refraction, 65
coding delay, 199
colour weather radar, 163
communications, 36–43
comparison frequency, 205
cone of confusion, 83
cone of no bearing, 45
cosecant beam, 158
critical angle, 24

data-link, 148
dead space, 25
Decca, 202–215
  chart, 204
  comparison frequency, 205
  errors, 213
  lane identification, 209
  range and accuracy, 213
decision height indicator, 171
differential correction (GPS), 226
distance measuring equipment (DME), 132–41
Doppler, 180–96
  aircraft aerial systems, 183, 187
  DF, 195
  drift measurement, 185
  errors, 194
  frequencies, 180
  setting up, 189
  up-dating, 194
  VORs, 89–91, 195
duct propagation, 29

emission, types of, 12, 53

fading, 27
fan markers, 118
flight management system, 143
flyback, 130
flyback voltage, 130
frequency, 2
frequency modulation, 10
frequency pairing, 111
fruiting, 147

garbling, 147
GCA (*see* ground controlled approach)
global navigation satellite system (GNSS), 221
global orbiting navigation satellite system (GLONASS), 222
global positioning system (GPS), 221
grass, 131
ground conductivity (VLF), 218
ground DF, 44, 195
ground proximity warning system (GPWS), 174–179
ground radars, 151–5
ground waves, 18, 28

hand-held microphones, 42
harmonics, 205
height ring, 166
hertz, 2

holding patterns, 60
homing, 57, 81–3, 96–8
hyperbolic systems, 197–201, 204

ICAO operational performance categories, 112
ILS (*see* instrument landing system)
instrument landing system, 102–118
  facility performance categories, 112
  false glide path, 114
  frequencies, 110
  glide path transmitter, 104
  localiser transmitter, 102
  marker beacons, 108
inversion, 29
ionisation, 22
ionosphere, 21
iso-echo display, 163

Janus aerial, 187

keying, 9

lane identification, 209, 218
legislation (radio), 41
limacon, 74
lock-follow, 134
loop, 48, 68
Loran, 199–201
  accuracy, 201
  operation, 200

mapping radar, 157–60, 165
marker beacons, 108
memory, 191
microphones, hand-held, 42
microsecond, 119
microwave landing system (MLS), 116–18
millisecond, 119
modes, 142, 176
mode S, 148
modulation, 8–12
modulator, 16
multi-hop refraction, 26

NDBs (*see* non-directional beacons)
night effect, 65
noise, 31, 52, 69
non-directional beacons, 48–71
  factors affecting range, 52, 64
  frequency band, 53
  letdown, 62
  rated coverage, 52
  types of emission, 12

OBS (*see* omni-bearing selector)
Omega, 216–20
omni-bearing selector, 77
one in sixty rule, 88, 99
optical distance, 28, 29
oscillator, 16

permittivity, relative (ε), 29
phase, 5
polar diagram, 8
polarisation, 7
precision approach radar (PAR), 151
precision instrument runway, 102
primary radar, 121
propagation anomaly, 29, 218
protection range, 66, 86, 87
protection ratio, 66
protective altitude, 66, 86, 87
proximate traffic, 232
pulse, 120
pulse modulation altimeters, 172
pulse recurrence frequency, 120
pulse recurrence period, 120

QDL, QDM, QDR, QTE, QTF, QUJ, 44
QGH, 46
quadrantal error, 68

radar
  airborne weather, 156–68
  anode system, 128
  basic, 119–31
  cloud warning, 156–68
  ground, 151
  iso-echo display, 163
  primary, 121
  secondary, 126
radial, 75
radio altimeter, 169–73
  pulse modulation type, 172
radio magnetic indicator (RMI), 92–101
radio waves, 1–35
  amplitude modulation, 9
  duct propagation, 29
  frequency modulation, 10
  ground wave, 18, 28
  modulation, 8–12
  phase, 5
  propagation, 1, 16–35
  skywave, 21
  space wave, 28
  theory of propagation, 1, 16–35
reference point (ILS), 111
refraction, 22, 65

refraction, multi-hop, 26
relative permittivity, 29
resolution advisory, 230
R-Nav (*see* area navigation)

satcom, 41
satellite navigation, 221–9
saw-tooth voltage, 130
secondary radar, 126
secondary surveillance radar (SSR), 39, 145–50
selcal, 39
selective availability (SA), 223, 226
sidebands, 11
skip distance, 25
skywave, 21
slant range, 136, 159
space wave, 28
spectrum, 17
squawk, 146
static, 30, 67
strobe, 130
super-refraction, 29
surface waves, 18, 28
surveillance radar approaches, 151, 154

thunderstorms, 157, 163, 165, 179
time base, 129–130
traffic advisory, 230
traffic collision avoidance system (TCAS), 150
transmitter, 16
transponder, 132, 145
tropospheric scatter, 27

up-dating Doppler, 194

very low frequency (VLF) systems, 219–20
VHF omni-directional radio range, 72–91
  aggregate errors, 87
  bearings, 83
  designated operational coverage (DOC), 82, 95
  directional signal, 75
  frequencies, 80
  homing procedures, 82
  phase measurement, 74
  principle, 72
  range, 85
  reference signal, 72
  uses, 80
VOR (*see* VHF omni-directional radio range)

warning (GPWS), 174, 179
wavelength, 2
weather radar, 151, 156
windshear, 179

zone (Decca), 207